Shallowing Impact

Practical Pathways to a Sustainable Future Beyond Utopia

Dr. Viswanadha Reddy Maddi

©*Shallowing Impact: Practical Pathways To A Sustainable Future Beyond*
Dr. Viswanadha Reddy Maddi
1st Edition

Publication Date: June 2025
Price: ₹400/-| $15
ISBN: *978-93-7164-298-9*
Published by:
Adhyyan Books
Office No. 589,
Opposite Vivanta by Taj,
DDA SFS. Pocket-1, Dwarka,
Sec-22, New Delhi-110077
Website: http://adhyyanbooks.com
E-mail: contact@adhyyanbooks.com

Printed at: Repro Printers, Delhi

Disclaimer: This book contains illustrative examples, case studies, and data points that may be fictional or composite. While based on research and real-world trends, specific statistics and examples are for educational purposes only.

Dedicated to my family

About This Book

Shallowing Impact: Practical Pathways to a Sustainable Future Beyond Utopia provides a refreshingly pragmatic vision for environmental living that goes beyond buzzwords and guilt-inducing messages. Dr. Viswanadha Reddy introduces sustainability as a realistic, fulfilling lifestyle that adds to personal well-being as well as global health.

This in-depth guide takes theoretical environmental ideas and puts them into practical everyday habits. From addressing the psychology of overuse and halting wasteful behavior to conserving energy wisely and targeting hidden toxins such as micro-plastics, each chapter sets out practical solutions that readers can use right away.

The strength of the book is that it is a realistic outlook—accepting that perfection is not the aim, but intelligent progress is. Reddy illustrates how incremental, persistent changes to our material purchases, energy consumption, travel behavior, and online habits can produce outstanding collective effects. Examples from real-life situations from cities and towns throughout India and the world show how sustainable living changed districts and cities.

Instead of sermonizing on deprivation, this book opens eyes to how mindful decisions frequently result in cost savings, better health, and deeper community relationships. Each chapter concludes with practical call-to-action sections, providing readers with clear steps to take action for sustainability today.

A Sustainable World demonstrates that sustainability and high technology are not mutually exclusive, thereby presenting an invaluable resource for anyone wishing to live more mindfully without sacrificing quality of life.

Introduction

The term sustainability has turned into a cliché and a buzzword. It is prevalent in every aspect of life around the globe. Some of the minor measures that could significantly contribute to managing a sustainable environment are often overlooked. For instance, instead of printing receipts at ATMs, we could receive all the necessary details through messages on our mobile devices. At restaurants, and with food deliveries, it has become common to provide packs of plastic cutlery, including forks and knives. When placing an online order, there should be an option to select the required cutlery rather than automatically including it. Moreover, the use of toothpaste containing micro-plastics should be addressed, as these materials end up going down the drain.

Materialism-there has been a trend in society to overuse materials such as clothes, furniture, vehicles, etc. For instance, some people, if not all, buy clothes and vehicles, viz., cars, motorbikes, and furniture, etc., even if they already have them, and use the new ones once in a blue moon. This causes unnecessary spending of finances on the one hand and causes a shortage of space in terms of parking and traffic-related issues when we speak of vehicles.

Attitudinal change, there is a need for attitudinal changes in so far as waste management is concerned. Things such as throwing cigarette butts, wrappers, and empty cartons of edible items all around, though they seem to be small, can play a vital role if we put them in the bins that they are meant for.

Energy management, household electrical appliances such as exhaust fans, lamps, and fans, not turning off the electrical plugs after getting the batteries of the mobiles charged, etc., also play a resource crunch in terms of energy.

Sound pollution can be brought down to a significant level if the motorists do not blow their horns unless it is required. Blowing horns only if required has to be made a part and parcel of our lives.

Micro-plastics, toothpastes, cutting off extreme ends of plastic packs such as milk packs, etc., can be avoided by careful practice of waste management by avoiding using toothpastes that contain micro-plastics and by putting the cut ends of the packs in the proper waste bin.

Online food delivery, waste of financial resources, plastic waste and unnecessary spending of money can be avoided. For instance, plastic waste, i.e., plastic forks, spoons, etc., can be avoided by having an option online to know as to whether this plastic cutlery is required at all. If it is required, how many in quality? Instead of this, we can observe that some quantity of plastic cutlery is provided while delivering food. The excess plastic material will go in dust bins across the globe. This can be avoided by providing an option to request the quantity. Apart from this, small sachets of ketchup, mayonnaise etc. are put along with the food delivery parcels. I have observed that some people do not consume them at all and throw them in the garbage bin. This is nothing but throwing money in the garbage. This can also be avoided by providing an option to choose from.

Table of Contents

Redefining Sustainability Beyond Clichés **1**

From niche to necessity, sustainability's evolution

Small Actions, Huge Impact: The Potential of Everyday Decisions

Digital Solutions for Physical Waste (e.g., E-receipts vs. Paper)

The Psychology of Sustainable Decision-Making

Establishing a Personal Sustainability Ethic

Debunking Sustainability Myths and Misconceptions

Measuring Impact: How to Know Your Changes Matter

The Material Paradox **27**

The Psychology Behind Overconsumption

Fast Fashion's Environmental Footprint

Planned Obsolescence: Ending the Cycle

Case Study: Minimalist Living in Urban Environments

Taking a "Use What You Have" Attitude

The Real Cost of Ownership: Overlooking Purchase Price

Models of Shared Economy Designed to Minimise Material Waste

The Mindset Shift: Attitudinal Transformation 57

The Collective Impact of Individual Actions

Breaking Habitual Waste Behaviours

The Social Psychology of Environmental Responsibility

Teaching Sustainability Across Generations

Case Study: Communities That Transformed Through Attitudinal Shifts

The Role of Visual Cues in Behaviour Change

From Guilt to Empowerment: Positive Environmental Psychology

Energy Intelligence for Everyday Life 72

The Hidden Energy Drains in Modern Homes

Smart Technology and Energy Conservation

Renewable Energy on a Personal Scale

The Economic Benefits of Energy Efficiency

Beyond Electricity: Comprehensive Energy Management

Saving Resources Saves Money: The Dual Benefit

Building Wealth Through Conscious Consumption

Budgeting for Sustainability: Short-Term Costs vs. Long-Term Savings

The Hidden Subsidies of Unsustainable Practices

Sustainability in the Digital Age **142**

Technological Solutions to Resource Management

The Environmental Impact of Digital Life

Apps and Tools for Sustainable Living

Social Media as a Force for Environmental Change

Case Study: Tech Companies Leading in Sustainability

The Carbon Footprint of Cloud Computing and Digital Services

Utilizing Data Analytics to Enhance Personal Sustainability

From Individual to Collective: Scaling Sustainable Solutions **157**

Community-Based Sustainability Initiatives

Corporate Responsibility and Consumer Advocacy

Global Networks, Local Impact

Intergenerational Equity and Long-Term Thinking

Building Resilient Systems for Environmental Uncertainty

Conclusion: A Sustainable World Within Reach **171**

Chapter 1

Redefining Sustainability Beyond Cliches

From niche to necessity, sustainability's evolution

Recall when "going green" was a topic of discussion solely among a tiny handful of activists? Not too far back was it. Environmental issues were sometimes considered as peripheral problems in the 1970s when Earth Day was observed. Today, sustainability appears in corporate boardrooms, classroom environments, and dinner tables all around, having become a regular topic of discussion.

This trip from niche to need did not occur over night. The concept of sustainability has long ago origins. The Brundtland Report published by the United Nations in 1987 helped to define "sustainable development" as "development that meets the needs of the present without compromising the ability of future generations to meet their own needs." This modest but significant sentence underlined that sustainability is about balancing the health of the ecosystem with people's needs and economic development, not only about the planet.

Since then, we have discovered quite more. According to the Global Environment Outlook series of studies published by the United Nations Environment Program (UNEP), sustainability is all around us, connected to everything we do from what we eat, to how we plan our cities. According to the 2019 Global Sustainable Development Report, sustainability is now absolutely required for our existence and success; it is not a choice.

But something intriguing has occurred as sustainability has become more mainstream. The term itself has occasionally lost significance and become what many refer to as a “buzzword.”

So many feel disconnected. They’re constantly hearing about sustainability but don’t know what it actually translates to in their day-to-day lives. A Globe Scan survey in 2021 discovered that 74% of global consumers are eager to live more sustainably, yet most feel uncertain how to do this.

This is where we start, stealing back the meaning of sustainability from buzzword territory and restoring it to something practical, personal, and powerful. True sustainability isn’t about being perfect. It’s not about shelling out big bucks for eco-stuff or changing your whole life overnight. Rather, it’s about seeing how our daily decisions impact the world around us and making small, lowering your carbon footprint. You're also getting healthier, possibly saving money, conserving traffic congestion, and influencing people around you that non-motorized transportation is possible. Your conspicuous action may encourage neighbors, friends, or relatives to do the same.

Well-documented is this "social contagion" effect. According to a 2018 Nature Human Behaviour study, people are much more willing to install solar panels themselves when they witness their neighbors doing so. This is the "neighbour effect," according to the researchers, and it implies that sustainable practices might proliferate through societies just as fashion or social trends would.

Massively done, even small actions can have a significant impact. The World Wildlife Fund's Living Planet Report notes that if everyone turned off the tap while brushing their teeth, for example, the global water savings would be astounding, billions of gallons daily. And the energy savings from everyone switching to LED light bulbs would be like millions of vehicles off the road.

The lesson here is that sustainability is about many people making more conscientious decisions every day, not about grand gestures or perfect fixes. "Never doubt that a small group of thoughtful, committed citizens can change the world; indeed, it's the only thing that ever has," said the anthropologist Margaret Mead famously.

Therefore, remember that you are a part of something so much more the next time you feel your acts don't count. Every sustainable action you do creates waves that mix with other people's choices to produce significant good change. In sustainability, small is very beautiful.

Digital Solutions for Physical Waste (e.g., E-receipts vs. Paper)

In today's more networked world, digital technologies have some of our most available options for lowering physical waste. Not only are these digital substitutes convenient, but they are also an enormous opportunity to shed our environmental burden without much change in lifestyle.

Take paper receipts, those tiny pieces of paper that pile up in wallets, purses, and ultimately garbage cans. The United States alone consumes around 10 billion pounds of thermal paper receipts per year, Green America's "Skip the Slip" report estimates. That takes about 10 million trees and 21 billion gallons of water to make. The environmental price tag doesn't stop there, most thermal paper receipts are coated with BPA or BPS, chemicals that can taint recycling streams and be a health risk.

The online option,e-receipts,is the easy way out. According to a 2019 Journal of Industrial Ecology study, e-receipts lower paper usage by 100% and also cut carbon emissions of receipt making up to 40%. Best Buy, Apple, and Whole Foods are some large retailers that already provide e-receipts, and banking apps for many institutions automatically categorize digital receipts, making them more convenient than keeping paper receipts.

India's digital payment revolution (UPI, BHIM, etc.) which has reduced paper receipts and physical currency use while increasing financial inclusion.

The rule applies much further than shopping receipts. Think through these common uses for digital replacement:

Bank statements and bills: The Environmental Paper Network estimates that if only 20% of American homes went electronic on bills, statements, and payments, we would eliminate around 151 million pounds of paper per year, save 1.8 million trees, and lower greenhouse gas emissions by 2.1 million tons.

China's adoption of QR code payments that has nearly eliminated paper receipts in urban areas.

Books and magazines: While nothing truly takes the place of the touch of a book for most readers, e-books have environmental benefits for avid readers. According to a 2009 study by the Cleantech Group, e-readers can pay for themselves in terms of reduced carbon footprint after only 20 books, with a full life-cycle analysis of digital devices and printed books.

Office documents: The Paperless Project alliance estimates that the typical office worker consumes approximately 10,000 sheets of paper per year. Cloud-based document sharing and digital signature tools can cut this usage dramatically. Paperless companies cite not only environmental advantages but frequently enhanced workflow efficiency and document security.

Even everyday communication provides space for digital substitution. Christmas cards, invitations, and friendly messages, historically delivered on paper, now come with digital equivalents that minimize waste while sometimes also offering improved functionality such as interactive features or multimedia content.

Nonetheless, it is worth noting that digital alternatives do have their environmental implications. Digital devices take resources to produce, use energy in operation, and generate e-waste when their life comes to an end. The Shift Project, a think tank specializing in carbon transition, estimates digital technologies account for around 4% of greenhouse gas emissions globally.

The solution to reaping the most out of digital alternatives is careful implementation:

- Utilize existing equipment instead of buying new ones dedicated to digital substitutes
- Maintain the life of your devices by maintaining and repairing them properly
- Make proper recycling of electronic devices at the end of their life
- Select cloud computing driven by renewable energy whenever possible

The digital revolution provides us with highly effective means of minimizing physical waste, but as with any tool, the outcome depends on how we utilize them. By exercising self-aware choice in deciding which digital alternatives work for our lives and responsibly applying digital technologies, we can greatly cut down our material footprint without sacrificing or even increasing convenience.

The Psychology of Sustainable Decision-Making

Why do some sustainable practices come easily and others appear almost impossible to change? The response is not simply in environmental consciousness but in the intriguing science of human choice.

As reported in a study published in the journal

Nature Climate Change, environmental awareness by itself does not often lead to sustainable action. The research discovered that although 70% of individuals reported concern for the environment, significantly fewer made environmentally conscious decisions on a regular basis. This "value-action gap" has confounded scientists for decades.

Recent breakthroughs in behavioral psychology are giving us some answers. In their seminal work "Nudge," Nobel Prize-winning economist Richard Thaler and legal scholar Cass Sunstein describe how humans do not always act on rational analysis. Rather, we use mental shortcuts, are guided by social norms, and prefer short-term rewards to long-term gains.

These psychological patterns create both challenges and opportunities for sustainable living. Here are some key insights that can help us make better environmental choices:

Default bias influences our behaviour more than we know. A 2020 study published in the journal Behavioral Science & Policy discovered that when renewable energy was made the default choice for utility customers (with an opt-out to select fossil fuels), participation in green energy programs jumped by 800% versus when customers had to opt in. This indicates that making sustainability the simpler choice, instead of taking more effort, can significantly boost adoption.

Social conventions have strong power. A study at Arizona State University showed that when guests at hotels were informed that most other guests reused their towels, there was a 26% increase in their towel reuse as well. We tend to follow automatically what we think others are doing, particularly the people who are like us or whom we respect.

Immediate feedback maintains new behaviors. An experiment in the journal Energy showed that homes that had real-time electricity monitors decreased their energy usage by 9% on average. The instantaneous feedback established a more direct linkage between behaviour and outcome than month-end utility statements could.

Loss aversion influences our choices. Researchers have discovered that we are more sensitive to losses than to equal gains. This is the reason why messages framing in terms of loss avoidance ("Don't lose the beauty of our natural parks") tend to be more powerful than messages framed in terms of gains ("Help make our parks more beautiful").

Our self-concept informs our decisions. As research in the Journal of Environmental Psychology found, individuals who define themselves as "environmentally conscious" tend to make environmentally friendly decisions in a variety of life domains. Embedding environmental values in our sense of self instills coherence across behaviors.

Knowing these psychological mechanisms is not the same as manipulating ourselves or others. Instead, it helps us to structure our environments and habits so that they facilitate rather than oppose the sustainable behaviour we wish to adopt. For instance:

- We can position recycling containers in obvious, easy-to-reach locations
- Install automatic programs for conserving energy, such as programmable thermostats
- Connect with groups where sustainable actions are the standard
- Structure immediate positive reinforcement for green decisions
- Frame sustainable choices in terms of what we gain instead of what we lose

By engaging with our psychological biases instead of fighting against them, we can make sustainable living less about constant effort and more about second nature. The best sustainable behaviors aren't ones that demand ongoing willpower but ones that ultimately become habits, things we do without much thinking.

As behavioral scientist BJ Fogg points out in his work on how habits form, the most long-term behaviors are those that become habits through repeated practice in favourable settings. If we know the psychology behind our decisions, we can set things up so that doing the green thing is the easy thing, the default thing, and ultimately, the thing we do without thinking.

Establishing a Personal Sustainability Ethic

Sustainability is strongest when it moves from being something we periodically consider to becoming an inherent reflection of our values and identity. Establishing a personal sustainability ethic involves creating an internal compass that directs our decisions, big and small, towards more environmentally and socially conscious outcomes.

Differing from strict adherence to outside rules, an individual sustainability ethic develops from your own values, situation, and knowledge of the environment. Such individuals, found in research through the Journal of Environmental Psychology, make more coherent sustainable decisions within various aspects of life even if no one sees them.

How then can you start developing your own sustainability ethic?

Begin with your "why." A study at the University of Rochester discovered that individuals tend to sustain ecologically friendly practices when driven by intrinsic values (such as connection to nature or care for generations to come) instead of rewards or social influence. Ask yourself: What elements of the natural world do I most cherish? Whose future do I care about? What type of world do I want to contribute to?

Direct experiences in nature, for many people, deepen environmental ideals. Time spent in natural environments boosts both emotional connection to the environment and sustainable behaviour, according to a 2020 study in the Journal of Environmental Psychology. These encounters, from gardening to forest hiking to just sitting by a nearby stream, can help you develop your own relationship to the natural environment.

Your sustainability ethic has to also be customized for your own circumstances and capacity. The 2018 IPCC Special Report on Global Warming underlined that we need different strategies for sustainability considering different settings and constraints. Living in a rural area without public transportation will give one different possibility for sustainable living than living in an urban apartment. Parent of young children will have different priorities than a college student or retiree.

This kind benefits rather than burdens. Inspired generations of environmentalists, environmental philosopher Aldo Leopold's "land ethic" holds that knowledge and experience together create an ecological conscience. Your own unique perspective, based on your culture, location, career, avocations, and life experience, may help others to see sustainability possibilities they might not otherwise notice.

As you form your own personal sustainability ethic, heed these guiding principles:

Sufficiency: The idea of "enough" is strong in a world that tends to promote excess. Studies by the Centre for Economic and Policy Research indicate that prioritizing sufficiency, having enough for a good life, rather than trying to optimize consumption, relates very strongly to both individual well-being and lower environmental footprint.

Systems thinking: In a 2021 Stanford Social Innovation Review article, it was stressed that realizing connections between ostensibly discrete problems assists us in making better decisions. For instance, recognizing how food decisions relate to water consumption, climate change, and biodiversity assists us in making more integrated choices.

Justice and equality: The 2019 UN Global Sustainable Development Report emphasized the point that sustainability must not be disentangled from social justice. An ethic of personal sustainability taking into account equity, such as making environmental costs and gains equitably accessible to all, is helping to secure more long-term solutions.

Regeneration: Instead of merely minimizing damage, most sustainability thinkers today focus on regenerative methods that enhance environmental health. The Regenerative Agriculture Initiative at California State University, Chico proposes that this attitude, seeking ways to leave things better than we found them, can revolutionize our relationship with the environment.

Ultimately, your personal sustainability ethic isn't something you perfect once and then follow rigidly. According to environmental educator David Orr, ecological literacy develops through lifelong learning and adaptation. Your understanding will evolve as you learn more about environmental issues, try new approaches, and discover what works in your life.

By developing this inner compass, you liberate yourself from the requirement of constant outside advice on what's "green enough." You build instead the ability to consider new circumstances, new technologies, and new choices in terms of your own thoughtfully considered values. A personal sustainability ethic doesn't make perfect decisions easy, but it makes reflective decisions feasible, and that's where real change starts.

Debunking Sustainability Myths and Misconceptions

Since sustainability went mainstream, there have been increased myths and misinformation along with sound facts. They discourage efforts, cause guilt trips, or give us hope of making change somewhere else in vain while the significant opportunity elapses. **Here are some common sustainability myths and the facts to believe**.

Myth 1: Our own actions will make little difference as opposed to pollution from industrial-scale activities.

Though it is the case that big industries play a major role in causing environmental issues, this does not imply that people's decisions are not important. As per Project Drawdown, which examines climate solutions, altering individual consumption directly propels about 25-30% of reductions required in emissions. Also, our consumers' choices affect companies' ways of working. The Nielsen Global Corporate Sustainability Report discovered that 73% of global consumers would absolutely alter their consumption patterns to minimize environmental impact, and businesses are responding to this trend more and more.

Myth 2: Being sustainable is too costly for the majority of individuals.

While some eco-friendly products carry premium

prices, many sustainable practices actually save money. The International Renewable Energy Agency's 2020 report found that energy efficiency improvements typically pay for themselves multiple times over during their lifespan. Similarly, a study in the Journal of Industrial Ecology found that households that reduced food waste saved an average of $370 annually. From water conservation to reduced consumption, many sustainable choices benefit both the planet and your wallet.

Take the example of Japan's "motional" cultural, principle that discourages waste.

Myth 3: Sustainability means sacrificing comfort and convenience.

This myth is based on outmoded visions of environmentalism as deprivation. Many of today's sustainable technologies actually improve, not decrease, quality of life. In J.D. Power's Electric Vehicle Experience Study, 82% of electric vehicle owners intend to buy another EV,partly because they like the driving experience, not merely because they like the environmental reasons. Likewise, energy-efficient residences generally provide better temperature control and air quality compared to less efficient structures. It is all about prioritizing the solutions that maintain or enhance quality of life at the expense of less environmental pressure.

Myth 4: Recycling solves the waste problem.

Although recycling has its uses, its limitations tend to be misunderstood. The EPA's 2020 figures indicate that merely 32% of municipal solid waste is recycled or composted in the U.S., and numerous products put in the recycling bin are not actually recyclable. Reduction and reuse tend to produce much greater environmental gains than recycling, the World Economic Forum says. A sustainable approach to materials prioritizes refusing unnecessary items, reducing consumption, reusing where possible, and only then recycling, followed by proper disposal as a last resort.

Myth 5: Sustainability is mainly about the environment.

The most enduring fallacy may be that sustainability only concerns the environment. In fact, the United Nations Sustainable Development Goals embody a far greater perspective that encompasses social justice, economic opportunity, and human well-being. The 2019 Global Sustainable Development Report highlights that environmental sustainability is not possible without confronting inequality, education, health, and the economy. Real sustainability is a balance in which human needs are provided for within planetary boundaries.

Myth 6: Complete perfection in sustainability is the goal.

Chasing the perfect green choice is the source of what psychologists' term "eco-anxiety" or "green fatigue." In a 2019 study published in the Journal of Environmental Psychology, individuals who were stressed by environmental perfectionism actually had less chance of long-term sustainable behaviour. Experts such as Katherine Hayhoe share that sustainability is a process of constant betterment and not necessarily the best place you can get to. The most sustainable path is one you can realistically keep up in the long run.

Myth 7: New technology will save us without changes to our way of life.

Though technology innovation is critical to sustainability, the International Energy Agency and other bodies continually report that technology alone cannot solve environmental issues without altering consumption patterns. The 2022 IPCC report on climate change mitigation reiterated that both technological and behavioral adjustments are necessary. A balanced strategy appreciates the importance of technology but realizes that how we live and what we cherish also needs to change.

By laying these myths to rest, we can concentrate on sustainability strategies that work and that are available to us. The reality is that sustainability has nothing to do with perfection, renunciation, or high-cost items. What it is about, fundamentally, is making deliberate choices that support the way things naturally are in nature and in our human experience, choices that are simpler and second-nature with use. Once we transition past myth to empirically informed awareness, sustainability no longer feels daunting but instead actionable in daily life.

Measuring Impact: How to Know Your Changes Matter

Perhaps one of the most frequent questions individuals have when implementing sustainable practices is quite simple: "Is this actually making a difference?" This need to know our impact is natural and vital. Without some means of measuring effects, it's hard to remain motivated or to know which actions are worthy of our limited energy and time.

Although flawless measurement is often not an option with intricate environmental systems, there are numerous practical methods of monitoring the effects of your sustainable behaviour. A study published in Environment International indicates that individuals who are given feedback regarding their environmental behaviour are much more likely to sustain and increase such habits over time.

Several methods of measuring your sustainability level have been discussed below:

Carbon footprint calculators are a good place to start for most individuals. Resources such as the EPA's Carbon Footprint Calculator or the Ecological Footprint Calculator from the Global Footprint Network estimate your environmental footprint based on your lifestyle. Not entirely accurate, these calculators can give you an idea of your baseline footprint and where to make high-leverage changes. 2021 research in Climatic Change revealed that individuals who utilized carbon calculators generally lowered their emissions by 14% compared to individuals who were provided with general information on sustainability.

In terms of energy consumption, your utility bill provides tangible readings. Most utility providers now include tools that monitor your electricity, gas, and water consumption over time so that you can tell if conservation efforts are paying off. Some even reveal how your use stacks up against comparable households. In studies conducted by Opover, those homes that received comparative feedback cut their energy usage by an average of 2-4% in the long term, a tiny fraction that, spread across a large number of households, equates to large sums of savings.

Water consumption can be similarly tracked through bills or meters. Installing a water meter on your main supply line or using smart water monitoring systems can provide detailed information about usage patterns. The Alliance for Water Efficiency reports that households that actively track water usage typically reduce consumption by 15% or more simply through increased awareness.

For waste minimization, even a basic weighing system can be of great benefit. In a study in the Journal of Environmental Psychology, households that weighed out their trash and recycling once a week for four weeks then decreased their waste output by an average of 20%. The numeric and visual feedback allowed them to see where opportunities for waste minimization lay.

Financial monitoring also offers indirect measurement of environmental effect. Most sustainable decisions, ranging from energy efficiency to decreased consumption, appear as savings in your wallet. Software like Mint or YNAB can track money spent on categories such as utilities, fuel, and consumer purchases, indirectly keeping tabs on your resource use.

Beyond individual measurements, community-based tracking can provide motivation and context. Organizations like Cool Climate Network have developed neighbourhood carbon footprint maps that show how your community compares to others. Participating in community challenges or sharing resource reduction goals with neighbors can provide social reinforcement for sustainable choices.

Online resources are making it easier to measure impact. Mobile apps such as Joule Bug, Green Choice, or Capture enable you to monitor daily sustainable behaviors and estimate their environmental value. Although the estimates these programs give are not exactly accurate, they provide good feedback on the relative impact of various options.

When direct measurement is not possible, proxy measures can still give valuable feedback. For instance, you can monitor how many meals you have that don't include animal products, how many days of commutes you do without a car, or how many months of no new clothes you have. Such behaviour counts can be useful proxies for the environment.

Don't forget that absolute measurement is not required for action that matters. Environmental psychologist Susan Clayton says consistency in method of tracking is more important than pure precision. No matter whether you are using complex carbon calculators or basic habit tracking, the important thing is to have a baseline and note changes over time.

While you're measuring your impact, keep things in perspective about what the numbers do and don't indicate. Environmental impact is seldom ever entirely good or bad, it's a degree of better or worse options based on your specific situation. It's not about getting a perfect score but about observing trends toward improvement over time and knowing which actions have the greatest positive impact in your specific scenario.

By discovering effective methods of measuring your impact, you turn sustainability from a theoretical concept into a tangible practice with visible outcomes. This feedback loop not only assists you in making better decisions but also gives you the satisfaction of observing how your individual actions lead to larger environmental solutions.

Call to Action:

Discover three "automatic" waste habits in your daily life and apply digital or low-waste alternatives this week.

As we finish this chapter, here's a quick two-step process to start your sustainability journey:

Step 1: Find three default waste habits

Observe habits in your daily life that generate unnecessary waste

Search for habits such as:

- Taking paper receipts when e-receipts are an option
- Using disposable products when reusable ones are available
- Leaving devices plugged in when not in use
- Printing out documents you could read digitally

Step 2: Adopt simple substitutes

For each habit, experiment with an easy, low-waste substitute

Examples are:

- Asking for e-receipts rather than paper
- Bringing a refillable water bottle or coffee cup
- Using a power strip that can be shut off
- Storing documents electronically rather than printing

What three habitual wasteful behaviours will you change this week?

Chapter 2

The Material Paradox

The Psychology Behind Over-consumption

Ever gone into a shop to purchase a single item and come out with a bag full of stuff you didn't plan on buying? Or caught yourself browsing online stores in the late evening, tapping "buy now" on products that crossed your mind an hour before? If this is you, then you're among many.

Our dynamic with possessions has grown complicated. Though possessing more than ever before, much of us harbour a nagging desire to continue buying more. A 2021 One Poll survey found the average American keeps around 300,000 products in their home, yet numerous still feel insufficient.

This consumption drive transcends needs or even rational desires. It gets at something more primal in our psychology. This has been examined for decades by marketing scholar Dr. Tim Kasser of Knox College. His research, appearing in the Journal of Consumer Psychology, indicates that most of us are shopping to satisfy emotional as opposed to material needs.

"When we're feeling insecure, anxious, or unhappy," Dr. Kasser continues, "turning to stuff can give us a fleeting feeling of control or happiness. But these emotional rewards are short-lived, leading to a cycle of consumption, momentary satisfaction, disappointment, and consumption again."

There are a number of psychological reasons that lead to over-consumption:

Social comparison is a strong force. When we notice friends, neighbors, or social media personalities with new devices, trendy outfits, or lovely houses, we can't help but compare our belongings to theirs. The American Psychological Association states that social media websites greatly enhance these comparison behaviors. Every time a friend shares a vacation picture from an upscale resort or flaunts a new acquisition, it quietly alters our definition of what's normal or essential.

Retailers recognize and take advantage of these psychological triggers. Observe how retailers generate artificial scarcity through "limited time offers" or "while supplies last" messages. These strategies trigger our fear of missing out (FOMO). In the same vein, loyalty programs, point systems, and tiered rewards create what psychologists refer to as "intermittent reinforcement", the same mechanism that makes gambling so addictive.

Even the store itself is structured to make us spend more money than we intended. That's why staples such as milk are usually found at the back of supermarkets, making you have to pass by scores of inviting products. The soothing light, enjoyable music, and tastefully laid out shelf displays all lower our resistance to spending.

Our brains also react to shopping in fascinating ways. Neuroscience studies employing brain scans indicate that the brain releases dopamine, a chemical linked to enjoyment and reward, when we expect to make a purchase. This occurs before we make the purchase. Alas, much of this pleasurable sensation wears off once we make the purchase, and we're left craving the next shopping "high."

Cultural messaging supports these tendencies. We learn from an early age that to be successful is to have more, a larger house, a new car, the latest gadgets. While earlier generations may have identified success in terms of community respect, family bonds, or handiwork, today's yardsticks tend to measure success by material accumulation.

Recognizing these psychological motivators doesn't leave us helpless in their grasp. The first step toward more intentional decision-making is awareness. If you catch yourself wanting to purchase something that isn't on your list, ask yourself:

- Am I buying to meet an emotional need?
- Is social comparison at the root of this want?
- Is marketing instilling false urgency?
- Will this purchase bring me lasting satisfaction?

Simple practices like waiting 24 hours before making unplanned purchases, keeping a wish list instead of buying immediately, or tracking your emotional state while shopping can help break automatic consumption patterns.

Remember, the goal isn't to stop enjoying material goods altogether. Rather, it's to bring awareness to our choices so that our purchases align with our genuine needs, values, and long-term wellbeing rather than psychological triggers carefully designed to separate us from our money.

Fast Fashion's Environmental Footprint

That adorable $15 impulse buy of a shirt last month, the one currently hanging untouched in your closet, has a history that reaches around the world and affects our collective environment in surprisingly large ways.

Welcome to the realm of fast fashion, fashion created to be cheap, on-trend, and, importantly, disposable. Whereas earlier generations may have bought new clothes for a season or for a party, the current fashion system now runs on one of perpetual newness, with some companies rolling out up to 52 “micro-seasons” annually based on a study by the Ellen MacArthur Foundation.

This perpetual churn has changed our relationship with clothing. Decades ago, individuals generally maintained clothing for years. Now, the average piece is worn only seven times before it is discarded, based on a WRAP study. In China, a survey of young women determined they find a piece of clothing “old” if it is only worn three times.

Textile waste in Panipat, India (known as the “cast-off capital”) where tons of Western clothing end up being sorted and recycled.

The effects on the environment are eye-opening:

Water use is perhaps the most shocking. It takes around 700 gallons of water to make a single cotton t-shirt, according to the World Resources Institute, approximately what one person consumes in 2.5 years. Jeans are even worse, taking about 2,000 gallons per pair. That water use takes place mostly in areas already facing water stress. In Uzbekistan, for instance, cotton production has led to the virtual vanishing of the Aral Sea, which was once the world’s fourth-largest lake.

The carbon footprint goes far beyond transport.

The fashion sector generates around 10% of the world's carbon emissions, as estimated by the UN Environment Programme, higher than international flights and shipping combined. This is due to energy-intensive production processes, the manufacture of synthetic fibres, and agricultural effects from cultivating natural fibres.

Another severe effect is that of chemical contamination. The textile dyeing business is the second-largest water pollutant in the world. Where there are relaxed environmental regulations in other nations, effluent wastes from dyes and chemicals of treatment pour untapped into country water systems. A Greenpeace documentary followed what happens to residents around textile production plants with this type of pollutant, presenting high incidence statistics of certain cancer diseases and infant abnormalities.

The textile industry along rivers in Bangladesh and India has created severe water pollution crises. Untreated chemical effluent from dyeing factories pours directly into rivers like the Ganges and Buriganga, turning waters into toxic flows while devastating aquatic ecosystems and threatening the health of millions who depend on these waters for drinking, farming, and daily use.

The waste problem is also troubling. Americans alone discard approximately 80 pounds of clothing per person every year, with approximately 85%

going to landfills or being burned. Synthetic fibres such as polyester, currently the most typical fabric used in clothing, take centuries to break down.

You may ask: "And what about giving away my spare clothes? Doesn't that rectify the issue?" Alas, the amount of discarded clothing has saturated the second-hand industry. The CBC Marketplace report "The Dirty Secret" found that only some 20% of donated clothing is sold locally. A lot of the remainder goes to developing nations, where it can upset native textile industries or, more and more, find itself in landfills there instead.

The social cost of fast fashion gives a further dimension to its real environmental cost. Fashion workers, largely women, are frequently employed in hazardous conditions at wages that prevent them from lifting themselves out of poverty. The 2013 Rana Plaza factory collapse in Bangladesh, killing more than 1,100 workers, made global headlines about these problems, but change continues to come slowly.

- So, what can we do? While systems change is finally necessary, individual actions still have an impact:

- Buy less, choose well. Try adopting a "one in, one out" policy in your wardrobe, or see how long you can last without shopping for new clothes for a month.

- Seek out durability rather than trendiness. More expensive products might be a higher upfront cost but often they will last longer, saving you money and resources in the end.

- Research the brands before purchase. Good On Your apps grade fashion brands according to their environmental and ethical strategies.

Knowing how to mend minor damage can give your clothes a lot more life. A basic sewing kit and some online tutorials can enable you to make simple repairs rather than throwing away otherwise serviceable items.

As shoppers, our purchases send a strong message to the market. In abandoning the use of fashion as disposable and moving toward more careful consideration in clothing, not only do we cut our individual environmental impact, but we encourage the industry toward greener and more sustainable directions.

Planned Obsolescence: Ending the Cycle

Why did your printer just fail right after its warranty period had ended? Why can't you change the battery in your cell phone? Why don't they produce refrigerators that will last 30 years like your grandparents used to have?

These disgruntlements allude to an increasingly prevalent business strategy: Planned obsolescence involves intentionally designing things to break down, go out of style, or otherwise require replacement within a relatively short space of time.

Now, planned obsolescence is practiced in several ways:

Technical obsolescence refers to when goods are built to fail or break down after some time. Many printers were discovered by the French consumer protection agency to have page-counting chips that induce “failures” once preset thresholds are met.

Design obsolescence occurs when new models have cosmetic modifications that render previous versions to appear dated, even if functionally identical. Consider how fast a three-year-old smartphone begins to appear “old-fashioned” while doing pretty much the same thing.

Incompatibility obsolescence takes place when fresh accessories, software, or replacement parts are not compatible with earlier models. Notice how charging cables keep changing with new devices? Or how software updates cease to be available for older computers after some time?

Repair restriction is another widespread practice. In a report, the U.S. Public Interest Research

Group states that most manufacturers employ specialized screws, glued parts, or proprietary diagnostic equipment to render repairs impossible or extremely difficult for independent repair facilities or consumers. Consider smartphones with batteries embedded within the case or appliances in which a single damaged part requires the entire unit to be replaced.

These activities bring about considerable environmental impacts. E-waste has become the quickest-growing waste stream in the world, with more than 50 million metric tons produced each year based on the UN's Global E-waste Monitor. These wastes have recoverable materials like gold, silver, copper, and rare earth metals that should be recycled. Instead, it is mostly deposited in landfills.

The economic effect on consumers is also alarming. The European Environmental Bureau calculates that lengthening the life of smartphones and other devices by a single year would save EU consumers €12 billion each year and cut carbon emissions equivalent to removing 2 million cars from roads.

To their credit, there is growing momentum to end this cycle:

The Right to Repair movement is gaining momentum all over the world, with several regions considering or enacting laws. These laws are intended to make manufacturers give repair manuals, offer spare parts, and produce products that are repairable and not disposable.

A few businesses are embracing longevity as a marketing value. Brands such as Patagonia, Fairphone, and Framework Computer openly design items to last, be fixable, and be easily updated, demonstrating that business success is not dependent on endless replacement.

New business models such as product-as-service are coming into being, in which the producers' own products and hence are motivated to ensure that they have longer lifetimes. Philips, for instance, sells "lighting as a service" to commercial customers, being responsible for maintenance and upgrading instead of merely selling lighting fixtures.

We, as consumers, hold considerable power to shape these practices:

Research product lifespan prior to buying. Sites such as iFixit offer "repairability scores" for products, and Consumer Reports and others frequently incorporate durability into their product assessments.

Advocate for right-to-repair legislation in your state. Groups such as Repair.org report on pertinent policy efforts.

Acquire basic repair skills. Numerous quick fixes need only basically tools and knowledge available on the web.

Even if their items cost more initially, support businesses who create for lifetime and repairability. Usually, the long-term savings exceed the more expensive purchase price.

"The Earth is what we all have in common," environmentalist Wendell Berry once brilliantly noted. Choosing products built to last and avoiding planned obsolescence helps us protect not only our money but also the natural resources we share with all living entities on Earth.

Case Study: Minimalist Living in Urban Environments

Meet Sarah Chen, a 34-year-old Chicago marketing professional living in a 600-square-foot apartment. Her life five years ago was quite different. Her flat was roomier, with furniture, wall decor, and closets loaded with clothes many people never wore. She had more room and more possessions, but she was always stressed, more so, stressed about her mounting credit card costs of living this way.

"I was working more and more hours to be able to buy more things that I never did have time to use," Sarah notes. "It was like riding on an always accelerating treadmill."

Sarah's metamorphosis came about following time spent with a friend embracing minimalism. The room's seeming serenity and order really spoke to her. Most importantly, she saw her companion seemed to be less nervous and more attentive throughout their talks. Inspired, Sarah started researching minimalist living and made the decision to try it for herself.

Her initial move was tackling her closet. In line with the trendy "Kon Mari" approach, she took everything out of her closets and retained only those items that were useful or that brought her true happiness. This eliminated close to 60% of her wardrobe.

"I saw that I'd been spending money on a pretend life, party clothes for parties I never went to, work clothes for occasions I didn't need to go to, workout equipment for activities I never had time for," she says.

Then came the furniture and household goods. Sarah sold her dining table that had been used mostly as a mail collection surface and bought a smaller, multifunctional desk that was used for both work and meals. She gave away duplicate kitchenware, decorative items that needed dusting but were not sentimental, and books she had read but would never read again.

The biggest adjustment was when her lease expired. Instead of renewing, Sarah moved into a smaller apartment near her job. Although the square footage was smaller, careful planning made the space adequate for her requirements. The relocation lowered her rent by 30% and halved her commute time.

The fiscal effect was instantaneous. In a period of eight months, Sarah had paid off her credit card balances. She funnelled some of the savings into experiences such as cooking lessons and weekend getaways, things studies have determined that bring more sustained joy than buying stuff. She also boosted her retirement contributions, easing financial anxiety for what's to come.

Aside from financial gains, Sarah observed other changes. It only takes minutes now to clean her apartment compared to hours before. Mornings become less stressful as her streamlined wardrobe simplifies picking out clothes. With fewer belongings that require care and attention, she has more time for socializing, reading, and a community garden activity she participated in.

Sarah's experience illustrates key principles of urban minimalism:

Multi-purpose rooms and pieces of furniture make the most of available square space. Her living room is office, dining room, and entertainment area through careful planning and organization.

Proximity can be more important than space. By living nearer to work, public transportation, and conveniences such as grocery stores, Sarah does not often have to drive, saving on transportation costs and lowering her carbon footprint.

Digital minimalism is a complement to physical minimalism. Sarah scanned important documents and photos, keeping paper clutter at bay while still having access to what's important.

Quality rather than quantity is achievable with fewer things. "When I purchase something now, I can afford better quality because I'm purchasing much less in total," Sarah says. "These things last longer, work better, and often provide more satisfaction."

Communal resources substitute for individual ownership. Sarah accesses books and media at the local library, goes to a nearby park rather than keeping a private yard, and joins tool-sharing at a neighbourhood association.

It's worth mentioning that Sarah's take on minimalism is not deprivation or austere limitations. She still has a TV, adorns with meaningful art, and maintains collections that make her happy (handmade ceramics by local artists in her instance). The distinction lies in intentionality, each item has a function or brings authentic joy.

"Minimalism is not about having less," says Sarah.

"It's about creating space, literally and figuratively, for what really matters to you."

Sarah's story represents a growing trend in urban areas, where residents are questioning whether bigger spaces and more possessions truly lead to better lives. The environmental benefits of this lifestyle are significant, reduced consumption, less waste, lower energy use, but the personal benefits of reduced stress, increased financial security, and more free time are equally compelling.

Taking a "Use What You Have" Attitude

"Eat it up, wear it out, make it do, or do without."

Popularized during the Great Depression and World War II, this proverb catches a mindset that is becoming more and more important now. The "use what you have" strategy is somewhat different from the default setting in our society, "when in doubt, buy something new."

This kind of thinking is about innovation, ingenuity, and finding the pleasure that results from maximizing what we already own, not about deprivation.

Think about the kitchen, a prime site for extraneous buying. Many homes have certain devices used once or twice before being thrown to the back of a drawer, a banana slicer, an avocado pitter, a strawberry huller. A good knife and chopping board would manage all these chores in meantime. Challenge yourself to fix the problem with what you already have before splashing money on a new kitchen gadget. Is the job you are thinking about purchasing a food processor for handled by your current blender? Would you be better off using a fork rather than a specialist mixing tool?

The same idea governs household repairs and maintenance. DIY techniques are now more easily available because to community tool libraries and YouTube tutorials. Many home problems can be resolved with a basic tool set and some internet advice without involving experts or complete replacement of objects. Many people's initial thought when facing a slow-draining bathroom sink is to immediately call a plumber. However, watching a quick instructional video on how to remove and clean the trap can turn this into a simple ten-minute task that costs nothing while providing a useful skill for the future.

One can apply this idea even in study and enjoyment. Review your shelves for unopened volumes or go back to favorites that call for another look before deciding on new books. Local libraries have developed into community resource centers providing not only books but also movies, music, tools, and even infrequently used items like cake pans and board games instead of purchases.

For the “use what you have” strategy, clothing has especially great possibilities. Instead of shopping for fresh clothing, fashion blogger Emma Johnson recorded her year of “shopping her closet”. She found lost objects, experimented with novel combinations, and picked basic altering techniques that turned seldom-used pieces into favorites. “I ended the year with not just a smaller ecological footprint,” she wrote, “but a better awareness of my personal style and what actually makes me feel good in my clothes.”

This kind of thinking also permeates cuisine. A Natural Resources Defence Council study indicates that the typical American home wastes about thirty-two percent of the food it buys. Meal planning using items already in your cabinet or refrigerator not only cuts waste but also often inspires creativity. Apps like Super cook let you enter your current ingredients and get recipe ideas, therefore transforming possible food waste into mouthwatering cuisine.

This strategy even helps electronics and technology. Usually, the most recent smartphone model provides just slight enhancements over past versions. Try renewing your present one first when tempted by a new gadget; update the software, delete pointless files, replace the battery if at all possible, or add a new case for a psychological refresh free of environmental effect from producing a new device.

This kind of thinking has advantages other than only environmental ones. Psychological studies pointing to resourcefulness and problem-solving as sources of more pleasure than passive consumption is mounting. Repairing a lamp instead of replacing it gives you not just a working light but also knowledge, confidence, and a narrative to share.

The financial benefits are also really important. The average American spends $18,000 annually on non-essential purchases, things they later characterize as "wants" rather than "needs." Redirecting even a portion of this spending toward experiences, savings, or higher-quality essentials can drastically improve financial wellness.

Communities have developed to help this kind of thinking. With groups in hundreds of local areas all around, the Buy Nothing Project helps neighbours to share and gift goods. Repair Cafés gather volunteers with a range of expertise to assist others in fixing broken objects rather than throwing them away. These community-based strategies show that making use of our resources does not imply doing everything by ourselves.

Changing this perspective requires work. Start by following a basic rule: list all you already own that might be useful before purchasing anything new. Ask, "How could I solve this with what I already have?" instead of looking right away for something to buy when confronted with a problem. With experience, this method becomes not only a sustainability tactic but also a creative and satisfying tool.

Traditional water harvesting structures in Rajasthan, including johads (village ponds), step-wells, and khadins (runoff-catching embankments), have supported communities for generations in this desert state. These brilliant indigenous methods not only store valuable rainwater but also recharge aquifers, proving sustainable solutions developed through generations of learning to adapt to water scarcity.

The Real Cost of Ownership: Overlooking Purchase Price

Whether you're in a store considering a purchase, a $30,000 car, a $300,000 appliance, or a $30 shirt, the price tag tells only a small portion of the tale. The actual cost of ownership goes much beyond that first figure to include costs and effects sometimes invisible at the point of purchase.

Knowing this larger picture can change our choices on what to bring into our life. Let's investigate what transcends mere cost.

For many things, maintenance expenses build over time and sometimes surpass the original purchase price. Take a home printer costing $100. Replacement ink cartridges during their lives will probably run about $400 or more. Likewise, a low-cost lawn mower could appear like a bargain until you consider regular maintenance and earlier replacement. The average car, according to the Automobile Association of America, runs roughly $9,282 yearly in maintenance, repairs, gasoline, insurance, and depreciation, expenses mostly left out of sight when considering just monthly payments.

Our calculations still mostly ignore storage expenses. Every square foot of our houses used for storage of belongings has financial worth, a fraction of our rent or mortgage that basically covers housing for our belongings. In pricey metropolitan regions, this might be really significant. Using a 10x10 area mostly for storage and paying $2 per square foot in rent, that comes to $2,400 annually to house belongings. “Many people are effectively paying hundreds of dollars monthly to store items they rarely use and could replace for less than their storage costs,” adds professional organizer Marie Smith.

The most often disregarded element of ownership is time expenditures. Everything we own takes some of our time for purchase, upkeep, cleaning, organizing, and ultimately disposal. Divining the cost by your hourly pay will help you to determine the “time price” of goods, according to economist Tim Harford. Still, even this fails to adequately represent the continuous time commitment many items demand.

The environmental expenses run across the lifetime of a good. Apple's environmental studies show that manufacturing one smartphone produces roughly 60 kg of CO_2 emissions, equivalent of many days of car travel. Manufacturing also calls for resource extraction, usually from fragile environments. Use generates other effects by consuming energy. Disposal creates more difficulties, especially for electronic waste, which sometimes includes hazardous elements needing specific management.

Psychological expenses can also be really considerable. Studies at Princeton University revealed that messy surroundings lower our capacity for information processing and concentration. Every possession calls for little choices, keep or throw away? Clean or disregard? Either fix or replace. These micro-decisions produce what psychologists refer to as "cognitive load," psychologically stressing us in ways we seldom credit to our belongings.

End-of-life expenses cover not only disposal fees but also the time and often emotional labour required in choosing what to do with objects we no longer need. Many people say they feel bad about throwing perfectly good objects, which results in extended storage of unneeded goods and continuous mental and financial expenses.

Knowing these hidden expenses enables better decisions:

Think about whole ownership cost instead than just purchasing price. Research lifetime costs including energy use, maintenance, and estimated lifespan for large purchases include appliances or cars. Two ways to assist measure some of these long-term expenses are Energy Star ratings for appliances and fuel economy ratings for cars.

Especially value your time and space. Before bringing new items home, take into account not only your affordability of the purchase price but also your capacity of the space they will occupy and the time they will need over their lifetime in your house.

Acknowledge the freedom associated with non-ownership. Often the least expensive choice is not buying a less expensive variant but not owning anything at all. For products used seldom, services like car-sharing, tool libraries, and rental platforms provide access without the hidden expenses of ownership.

For purchases that call for durability and repairability, pay close attention better-quality products can have higher upfront costs, but over time overall ownership expenses usually are lower. Companies such as Consumer Reports offer data on product dependability, so enabling the identification of really affordable solutions.

Manufacturer Patagonia tackles this idea directly in their product descriptions, noting: "The most environmentally sustainable product is the one you already own." By considering the true cost of ownership, to our finances, time, psychological well-being, and the planet, we can make decisions that serve our actual needs while lowering needless material throughput in our life and economy.

Models of Shared Economy Designed to Minimise Material Waste

Are you required to have a power drill you use for fifteen minutes annually? Does each house on your block have a lawn mower sitting idle most of the week? When a neighbouring company has the identical thing accumulating dust, would it make sense for an office to buy seldom used equipment.

These concerns draw attention to a basic inefficiency in our ownership-based economy: many objects spend the great portion of their useful lives idle. According to a RAC Foundation analysis, the typical automobile is parked 95% of the time. Based on studies by the Ellen MacArthur Foundation, the average power drill lasts just 12 to 15 minutes during its lifetime. These idle resources reflect both personal economic waste as well as unwanted environmental harm from manufacturing goods with low relative value given their resource needs.

Focusing on access rather than ownership, shared economy models provide different paths. These models show several shapes:

Usually enabled by online platforms, community-based sharing functions through local networks. Now set in hundreds of cities, tool libraries let members borrow everything from kitchen appliances to power equipment for a little membership fee or occasionally for free. One of the first in the United States, the Berkeley Tool renting Library notes renting over 25,000 tools yearly, therefore avoiding thousands of duplicate purchases.

Organized mostly through social media, Buy Nothing groups establish hyper-local gift economies where neighbors freely exchange items, therefore lowering waste and fostering community relationships. These clubs, which have members reporting both social and environmental benefits from involvement, now find homes in thousands of towns all over.

Participating in her neighbourhood Buy Nothing group in Philadelphia, Maria Lopez says: "Last month I needed a food processor for a special recipe. Instead of buying one, I posted in our group and borrowed one from a neighbour two blocks away. I not only saved money but also met someone from my neighbourhood I'd never spoken with before."

Commercial sharing systems have started to show up in several industries. Members of car-sharing companies like Zipcar and Get around can access automobiles just when needed. Rather than calling for new hotel building, accommodation sharing on websites like Airbnb makes use of already existing homes. People can borrow everything from cameras to camping gear from neighbors using peer-to-peer rental sites like Fat Llama.

These sharing strategies have really large environmental advantages. Car-sharing members cut their transportation-related carbon emissions by up to 50%, mostly by driving less generally and utilizing public transportation more regularly, according to a study written up in the journal Environmental Innovation and Societal Transitions. Comparatively, studies by UC Berkeley's Transportation Sustainability Research Centre found that each shared automobile substitutes between 9 and 13 privately owned cars.

Public bicycle sharing initiatives in Mysore and community refrigerator movements in Bengaluru and Mumbai exemplify India's growing shared economy, where the former reduces carbon emissions while the latter tackles food waste by allowing residents to share excess food with those in need, combining environmental sustainability with social welfare.

Apart from the effects on the surroundings, sharing economy involvement benefits providers as well as users financially. AAA reports that the typical auto owner spends $9,282 a year on their vehicle. Sharing services can save thousands of dollars yearly for someone who wants sporadic but not daily use to a car by giving access to many vehicle kinds as needs vary.

Sharing underused resources gives providers fresh sources of income. Many Airbnb owners say that part-time letting their houses helps pay property taxes or mortgage expenses, therefore making housing more reasonably priced overall. Likewise, peer-to-peer car sharing programs let owners rent to neighbors when their automobiles would otherwise be idle, therefore helping to offset vehicle costs.

Benefits to communities go beyond just resource effectiveness. Studies conducted for the Journal of Consumer Research revealed that involvement in sharing economy projects strengthens community trust and social relationships. Creating chances for neighbours to engage around pragmatic needs helps to create social capital alongside resource efficiency.

Business models are changing to allow for common access as well. While giving client access, the product-as-service model preserves manufacturer ownership. Businesses like as Mud

Jeans provide "jeans subscriptions," whereby consumers lease instead of buy jeans and return them for recycling when they're ready for something different. Using similar subscription arrangements for office furniture, furniture startup IKEA is incentivizing robust, repairable design over planned obsolescence.

Think about these ideas to include sharing into your own life:

Start with occasionally used products. Before you buy anything new, consider community sharing possibilities and list anything you use seldom, specific tools, party supplies, seasonal equipment.

Search your local region for established sharing facilities ranging from car sharing programs to tool libraries. Local sustainability groups sometimes keep resource sharing directories.

Consider co-ownership with friends, family, or neighbour for basic but costly products. Establishing basic agreements regarding access timings and maintenance tasks would help this method to go without problems.

See whether a sharing alternative may satisfy your needs with less environmental and financial effect before making big purchases. Sometimes opting not to own at all is the wisest ownership choice.

"The most sustainable product is the one that's never produced because we've found ways to meet our needs with existing resources," environmental researcher Anne-Sophie Carbonnelle notes. By embracing shared access over individual ownership, we can greatly lower resource consumption while often gaining access to higher-quality goods than we could individually afford."

Call to Action

Before your next non-essential purchase, set a 7-day waiting time and note if you still need it following the wait in your journal.

Our next moves toward more conscious consumerism can be straightforward yet significant:

First: Stop before you buy

- Instead of acting right away when you want to purchase something non-essential, note it.
- Mark a calendar reminder seven days later to review this possible purchase.
- Stow the item in a "consideration list" instead of your shopping basket.

Second step: consider throughout the waiting time.

Simple questions regarding the object will help you.

- Exists anything I could have right now that would help with this?
- Realistically, how often will I utilize this?
- Where is going to be my storage?
- Whose issue am I actually trying to fix?

This week, what non-essential purchase will you put off?

Chapter 3

The Mindset Shift: Attitudinal Transformation

The Collective Impact of Individual Actions

"What's the point? My actions won't make a difference." This kind of thinking permeates discussions about environmental problems all around. After all, if big businesses keep polluting on an industrial level and you correctly sort your recyclables, what difference makes?

Although this sense of powerlessness is natural, it obscures a vital truth: individual efforts, when multiplied throughout populations and communities, have great collective impact.

Think about what 2019 saw in Delhi, India. Thousands of individual homeowners made little adjustments when air pollution hit hazardous levels: carpooling, cutting firecracker use during Diwali, and using odd-even vehicle guidelines. Air quality levels clearly improved after weeks.

Environmental sociologist Dr. Arun Agrawal refers to this as "the power of aggregated action." Based on his studies, he discovered that when just 25% of a society adopts a particular environmental habit, it typically sets a tipping point that speeds up more general acceptance. This "social contagion" effect means that your own behaviour generates waves much beyond your direct influence.

Consider calls for plastic bag bans. These developments started with individual consumers carrying reusable bags before they became laws in many regions. Their observable activities normalized new behaviors, therefore policy changes both feasible and successful. Originally scattered individual decisions, billions of plastic bags were finally diverted from oceans and landfills.

The arithmetic creates an interesting narrative. One person avoiding single-use plastic bottles for a year keeps about 167 bottles out of the garbage flow. Should 100,000 people choose the same path, there are 16.7 million bottles involved. Individual activities have a quite different scale.

Remember that your obvious behaviour gives others permission to follow. The individual observing you might also decline a plastic straw the next time. The neighbour who notices your compost bin might wonder how one should be started. Your decisions provide both ongoing cascade of influence and direct environmental advantages.

Breaking Habitual Waste Behaviours

Most of us do not deliberate in wasting. Waste instead occurs from regular actions below our awareness: the wrapper thrown reflexively into the trash instead of being recycled, the lights left on while leaving a room, the food forgotten in the refrigerator until it goes bad.

Changing these behaviors calls a knowledge of how they develop and endure. Behavioral scientist Charles Duhigg argues that habits comprise three elements: a cue, a routine, and a reward. The signal sets off the behaviour; the routine is the behaviour itself; the reward helps the conduct to be likely to be repeated.

The signal might be finishing your takeaway lunch, the routine is tossing all things in the general rubbish, and the reward is the ease of a quick cleanup. We must modify the practice keeping the same cue and reward in order to break this habit.

One first step is mindfulness. Just knowing about automatic habits opens a chance for modification. Try this: carry a little notebook one day and mark each occasion you throw away something. Just observation; no judgement here. The amount most individuals unintentionally toss away shocks most of them.

According to studies by environmental psychologist

Dr. Elke Weber, developing "implementation intentions", particular plans for when and how you'll act differently, "significantly increases success in breaking habits." Rather than a nebulous target like "I'll waste less," consider "When I order takeaway (cue), I'll separate recyclables from general waste (new routine) before disposal of them."

Visual reminders help too. Automatic behaviour patterns can be disrupted by a little note near the light switch, a sign on the trash can, or a reusable bag hanging by the door.

Replacing wasteful habits works better than simply trying to eliminate them. If you buy bottled water regularly for convenience, trying to "stop" most likely won't work. Rather, develop the habit of every morning filling a reusable bottle and carrying it with you.

Recall that modification of habits takes time. Studies at University College London revealed that habit development usually takes two to three months. Celebrate little successes and be gentle with yourself.

The Social Psychology of Environmental Responsibility

Whose definition of "right" or "wrong" environmental conduct guides us? While some people believe they personally should be in charge of lessening their environmental footprint, others do not regard this as their issue.

The intriguing field of social psychology – the study of how social factors shape our ideas, emotions, and actions – has the answers.

Fundamentally, environmental responsibility stems from our social identity, that of the groups we hang about and the values they represent. Social psychologist Nilan Nirmalaghan's studies reveal that those who strongly identify with environmentally sensitive groups are more likely to feel personally accountable for environmental problems.

This helps to explain why environmental actions sometimes cluster in communities, not only because of common resources like recycling programs, but also because social conventions greatly affect our sense of duty. You are more inclined to view composting as a personal obligation rather as an optional activity when your neighbours and friends do.

There are two flavors to social conventions: injunctive (what society accepts or disapproves of) and descriptive (what most people really do). Both shape our environmental actions. According to a landmark study by psychologist Robert Cialdini, guests of hotels were most inclined to re-use towels when told that most other guests did so (descriptive norm) and that reusing towels was socially sanctioned (injunctive norm).

Also influencing our feeling of environmental duty is seen fairness. Should we see others not contributing their share, we could wonder about our own efforts. This "why should I bother if no one else is?" kind of thinking can undercut environmental improvement. On the other hand, obvious group activity shows shared commitment and so strengthens individual accountability.

Distance, physical as well as psychological, affects our sense of duty as well. Problems that seem far-off, like melting arctic ice, engage us less than local ones, like water contamination of adjacent lakes. This helps to explain why local environmental challenges typically inspire the same degree of personal responsibility as global ones like climate change.

Fascinatingly, environmental responsibility rises in those who feel they have agency, that ability to influence results. When provided with particular, doable behavioural adjustments instead than overwhelming global issues, respondents exhibited more desire to engage personally in environmental action, according to a University of Exeter study.

Teaching Sustainability Across Generations

The environmental problems we confront now will need for answers spanning several decades.

Teaching sustainability is about developing attitudes, practices, and thinking patterns that will guide children's interaction with the natural world over their life, not only about exposing them to environmental problems.

Good sustainable education starts early. Studies reveal that young children, even three-year-olds, can understand fundamental environmental ideas as water conservation or rubbish sorting. These early encounters set the stage for eventual more advanced knowledge.

Still, our teaching of sustainability has great influence. Environmental educator David Sobel issues a warning against what he describes as "ecophobia", fear-based instruction stressing environmental disaster. "If we fill children with facts about environmental destruction before they've developed a loving connection with the natural world," Sobel says, "we create fear and disengagement rather than care and commitment."

Early environmental education should, instead, stress natural pleasant experiences. Studies repeatedly reveal that early environmental experiences significantly correlate with adult environmental commitment. Simple activities like gardening, visiting nearby parks, or seeing animals lay an emotional basis for more advanced environmental education.

Learning across generations runs both directions.

Children typically bring fresh sustainable ideas into homes while adults impart formal environmental knowledge to them. Children who taught about recycling at school effectively changed family behaviors at home, acting as "sustainability champions," according to a study in the Journal of Environmental Education.

New opportunities for environmental education abound from digital instruments. Through familiar technology, apps gamifying waste reduction, platforms linking students with worldwide peers confronting similar environmental concerns, and citizen science projects supporting actual research all engage young people.

Most crucially, perhaps, environmental education has to be hopeful. Young people must come to know that their activities count and that good change is achievable. Case studies of successful environmental projects, contact with sustainability entrepreneurs, and chances to apply their own ideas foster agency and hope.

Case Study: Communities That Transformed Through Attitudinal Shifts

Many considered it impossible when Kami katsu, a small town in Japan with roughly 1,500 people, set its zero-waste target in 2003. The community has meagre resources and no sophisticated waste management system. Still, it was dedicated to a common vision and ready to change everyday behaviour.

Kami katsu today directs approximately eighty percent of its garbage away from incinerators and landfills. From metal cans to paper cartons to particular kinds of plastic, residents classify garbage into 45 groups. The town's metamorphosis resulted from a significant change in community attitudes rather than from sophisticated technology or large financing.

"It was first time-consuming and inconvenient," says resident Hatsue Katayama. "But gradually, we started to view our waste differently. Now, before buying something, I automatically consider how I'll dispose of it. That thought process has become natural."

The community of Vengeri in Kerala, India, struggled with waste management as modern consumption patterns and single-use packaging replaced traditional methods of composting food waste and re-purposing items. By means of neighbourhood waste committees, residents resurrected and modernized traditional garbage management instead of waiting for government answers.

These committees changed attitudes rather than only oversaw garbage collecting. By means of house-to-house instruction, community seminars, and school projects, they reinterpreted trash management as a shared obligation rather than a government function. Vengeri's all-encompassing waste management system today comprises home composting, a community biogas plant, and a "zero-waste" mindset that penetrates daily decisions.

Since Kami katsu and Vengeri tackled mindsets rather than only behaviour, they both succeeded. They understood that sustainable waste management is a social as well as a technical one requiring a change in strongly rooted ideas about convenience, responsibility, and community.

"The most successful community transformations don't impose external models of sustainability. They rediscover and reinvent sustainable elements already present in local traditions, so making environmental responsibility feel like cultural continuity rather than foreign imposition," notes environmental sociologist Dr. Amita Baviskar.

The Role of Visual Cues in Behaviour Change

Have you ever found that you are more inclined to use a clearly labelled recycling bin than an unmarked one? Alternatively, how does a clearly unclean park seem to draw more trash than a clean one? These replies show how strongly visual signals shape our environmental actions, often without our conscious awareness.

Like silent but effective environmental cues, visual ones serve as they work as much of our behaviour is driven by contextual cues suggesting suitable actions, functioning below our awareness.

Think through what happened when Wesley Schultz, an environmental psychologist, investigated beach trash. Visible trash can beaches saw 40% less littering than those without. Simple visual indicators like bright footprints leading to bins helped to cut pollution by another 17%. Since they affected behaviour right at the point of choosing, these visual paths performed better than written anti-littering messages.

Environmental visual cues especially rely on colour in a really strong manner. Blue recycling bins routinely surpassed other colours in recycling compliance, according to University of British Columbia research, most likely because blue has strong psychological ties with recycling. Studies conducted in retail settings, meantime, suggest that green signs raise buyers' awareness of environmentally friendly products by up to 27%.

Additionally affecting behaviour is visual closeness. According to a University of Pennsylvania research, side-by-side recycling and garbage bin placement with good visual labelling improved appropriate sorting by 34% over separate bins. People are more likely to make good decisions when sustainable options are aesthetically pleasing right at the decision point.

Another strong visual trigger are images of nature.

Studies reveal that just viewing pictures of natural surroundings can inspire environmental values and boost sustainable practices. According to one study, office workers whose desks had natural photos cut paper use by 15% as compared to those in settings devoid of such images.

From Guilt to Empowerment: Positive Environmental Psychology

Environmental message has for decades mostly depended on guilt, fear, and shame. Melting ice caps, suffering animals, terrible scientific projections, these pictures and stories seek to shock us into response. Environmental psychologists have found, however, something unexpected: negative emotions typically cause disengagement rather than consistent action.

"When people feel overwhelmed by environmental guilt or fear, their psychological defenses activate," notes environmental psychologist pioneer Dr. Renee Lertzman. "They may intellectually admit the problem but emotionally distance from it, a response that protects their sense of well-being but prevents meaningful action."

This realization has fundamentally changed environmental psychology from guilt-based methods to empowerment-oriented techniques fostering hope, agency, and positive vision.

Studies show that whilst negative feelings like

guilt, shame, or fear correspond weaker with consistent environmental actions, good emotions like optimism, pride, and fulfillment do. This means characterizing environmental issues as challenges we have the collective ability to solve rather than discounting them.

Soluble-oriented communication is quite different. People's sense of efficacy, belief in their potential to change things, gives rise by up to 50% when they learn about environmental issues together with workable remedies. This efficacy view acts as a psychological link between concern and behaviour.

Honouring advancement, even little victories, gives impetus. Environmental groups that showcase successes usually inspire more consistent participation than those just focused on issues. These successes build positive reinforcement that keeps one committed even through unavoidable challenges.

The language we choose is really important. Framing environmental actions as "gaining" something (health, community, savings) rather than "losing" something (convenience, luxury) sets more positive connotations. Likewise, phrases like "cleaner," "healthier," and "smarter" inspire more favourable involvement than words like "less harmful" or "reducing damage."

Positive environmental psychology is substantially

enhanced by community connection. Those who feel part of a group striving toward common environmental goals report higher degrees of hope, dedication, and well-being than those who view their activities as lonely attempts. This social aspect turns environmental action from responsibility to genuine involvement.

Call to Action

Plan a local clean-up day emphasizing usually disregarded micro-plastics and small waste products.

Here's a useful approach to apply attitudinal transformation as we draw to close this chapter:

Step 1: Plan Your Clean-up

- Choose anywhere in your neighbourhood park, beach, stream, or roadway.
- Choose a time and date that neighbors would find reasonable.
- Prepare for gathering small waste items for water locations using fine-mesh nets or sieves; gloves and small containers for land locations.

Step 2: Engage Your Community

- Invite neighbors with flyers, social media groups, or community bulletin boards.

- Stress community building as much as environmental impact.
- Ask everyone to bring reusable water bottles and their own gloves.

Step 3: During the Event

- Start with a quick overview of the reasons behind small waste significance.
- Show how to spot and gather little waste products.
- Work in small teams to foster ties to your community.

Step 4: After Collection

- Sort and count gathered objects to show the impact.
- Share findings with both participants and the larger society.
- Talk about easy behaviors that might keep some objects out of the surroundings.

What sometimes disregarded trash problem in your area would gain from group attention?

Chapter 4

Energy Intelligence for Everyday Life

The Hidden Energy Drains in Modern Homes

You had a demanding day and just got home. The house is unoccupied however a number of lights are on. Though silent, the television beams from the corner. Your phone charger is hooked in, not with a phone. The air conditioner hums, cooling rooms nobody is using.

These apparently small energy drains reflect what energy experts refer to as "phantom load," or "vampire power" – electricity absorbed by appliances even in apparently dormant or off states. The Lawrence Berkeley National Laboratory estimates that in the typical house, these phantom loads account for about 10% of the household energy use.

First step in properly managing your energy is knowing where it goes. As the figure below shows, water heating and large appliances follow heating and cooling, which usually consumes most of household energy:

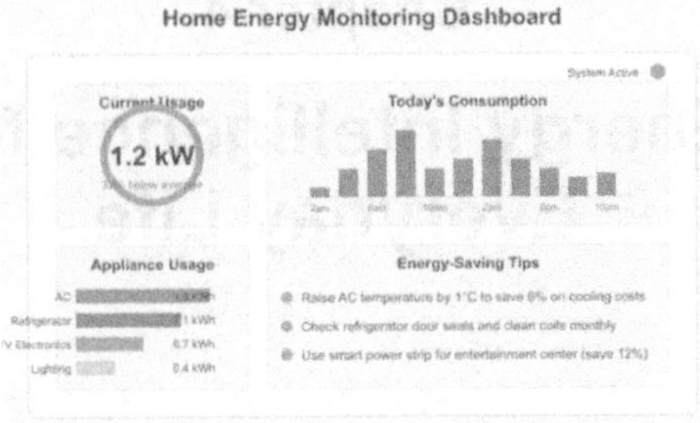

Source: U.S. Energy Information Administration (EIA), Residential Energy Consumption Survey (RECS). Chart showing the breakdown of energy consumption in typical residential buildings by end use category.

Think back on your entertainment setup. Many televisions, gaming consoles, and cable boxes keep running when "off," maintaining memory settings, updating software, or reacting to remote controls by sucking power. Although a normal TV uses 1,500 kWh yearly, up to 400 kWh happens while the gadget is ostensibly "off."

Similar dishonesty is found in kitchen appliances. Your microwave most certainly runs its digital clock on more electricity than it heats meals on. That harmless clock display uses roughly 35 kWh yearly, a little bit individually, but multiply it across your oven, coffee maker, and other appliances to soon add up.

Home offices are now major consumers of

electricity. Computers left in sleep mode rather than completely off can use 15 to 70 watts constantly. Along with this continuous drain, printers, monitors, and wireless routers contribute.

Heating and cooling systems represent the largest energy expenditure in most homes, accounting for nearly half of residential energy use. Still, inefficiencies abound: poorly sealed windows and doors, insufficient insulation, and thermostat settings that either cool or heat empty rooms all help to waste energy.

The excellent news is, once found, these latent energy drains are quite easy to treat. When not in use, power strips with switches can totally cut off electricity to homes offices and entertainment systems. When gadgets reach full battery, smart plugs can turn off chargers automatically. When none is home, programmed thermostats can cut heating and cooling.

Smart Technology and Energy Conservation

Recall when turning off a light meant for stepping across the room and flipping a switch? Smart home technologies of today have revolutionized energy management from human chores to automated systems optimizing consumption depending on real-time data and usage patterns.

Among the most important developments in home energy control are smart thermostats.

Unlike programmable thermostats, which run set schedules, smart thermostats over time understand your preferences and routines. Through smartphone apps, they may be managed remotely, sense when you're away, and adjust for changing conditions. Research by the United States Smart Thermostats, according to the Energy Star program of the Environmental Protection Agency, save an average of 8% on heating and 10% on cooling costs; or roughly ₹3,500-7,000 yearly for the ordinary Indian home.

Through motion sensors, daylight detection, and remote controls, smart lighting systems provide similar advantages. Lights in empty rooms or dim when enough natural light turns off automatically. Certain systems can even progressively change colour temperature over the day to mimic natural circadian rhythms, therefore enhancing sleep quality and perhaps lowering energy use.

Still, the modest smart meter is maybe the most effective smart technology available for energy savings. These devices turn an abstract idea into observable data by giving real-time energy consumption feedback. Studies repeatedly demonstrate that simply raised knowledge helps homes with energy meters cut consumption by 5-15%.

Energy researcher Dr. Veena Sharma says "When people can see their energy use in real-time, it creates an immediate connection between behaviour and consumption." "It's like the difference between watching money leave your wallet with each purchase and once-monthly bank balance check."

The advantages for the environment go beyond personal residences. More effective control of electricity networks is made possible as smart technology grows over towns and cities. Utility companies can use smart houses to momentarily lower non-essential system consumption during moments of great demand, therefore preventing blackouts and lowering the necessity for new power plants.

Renewable Energy on a Personal Scale

Many people's ideas of renewable energy centre on large solar farms or tall wind towers. For individual homes looking for greener power sources, however, renewable energy has also been declining and growing more readily available.

The most obvious sort of personal renewable energy available is rooftop solar panels. Originally somewhat costly, the cost of solar photovoltaic systems has dropped around 70% throughout the last ten years. For millions of homes, this huge price drop has made solar from a luxury for the environmentally conscious a sensible investment.

Typical rooftop systems in sunny parts of India such as Gujarat and Rajasthan can produce

4-6 kilowatt-hours per square meter daily. This translates to 70–90% of yearly power demands for a house with appropriate roof space. With many systems paying for themselves through utility savings within 5-8 years then delivering basically free electricity for decades thereafter, the economics have grown increasingly positive.

What about residences where rooftop solar is impractical, though? One such route to sustainable energy is provided by community solar installations. These shared systems let several homes make joint investments in and gain from a central solar system. Usually receiving credit on their utility bills commensurate with their investment, participants–renters, apartment residents, residences with inappropriate roofs–have access to renewable energy.

Many utility providers today provide green power buying choices for customers who cannot create their own renewable energy. These initiatives let consumers indicate, even if the actual electrons entering their houses come from conventional facilities, that a part or all of their electricity comes from renewable sources.

Most importantly, perhaps, personal renewable energy systems turn consumers into producers, that is, "prosumers" in business language. This change fosters higher awareness, usually results in more deliberate consumption practices, and establishes a more direct relationship to energy. Studies reveal that homes using solar systems usually cut their total energy consumption by 10–15%, independent of the renewable generating itself.

The Economic Benefits of Energy Efficiency

When one talks about energy efficiency, environmental advantages usually take the stage. Still, the financial benefits are equally appealing and usually inspire more quick desire for action.

Investments in energy efficiency can yield very amazing financial rewards. According to a thorough analysis by the International Energy Agency, increases in energy efficiency usually pay off annually between 25 and 50 percent, much above most conventional investments. Few other investments improve property prices and lower environmental impact while nevertheless providing such consistent, tax-free profits.

Not every energy efficiency improvement yields equal financial gains. The payback times and return on investment for typical energy efficiency upgrades are shown below:

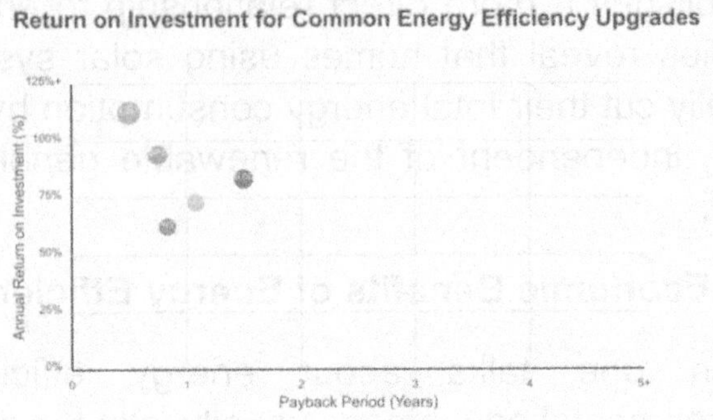

Source: U.S. Department of Energy, Office of Energy Efficiency & Renewable Energy, "Home Energy Efficiency ROI Analysis" (2021)

Think about lighting; among the easiest efficiency improvements is this one. Although it costs almost ₹2,000 to replace ten often used incandescent bulbs with LEDs, yearly electricity expenditures will be saved almost ₹3,000. With the investment repaid in just eight months, that is a 150% annual return. These bulbs will repeatedly pay for themselves throughout their 15-to-20-year lifetime.

Using 60% less energy than traditional fans, ceiling fans with BLDC (Brushless DC) motors provide still another high-return efficiency investment. Usually paying back the premium cost in two years, the annual savings for a home running fan several hours daily can approach ₹2,000 per fan.

Beyond simple cost savings, energy-efficient homes fetch top resale value. Energy-efficient homes sell for 2-8% more than equivalent properties with average efficiency, according to studies conducted across many property markets. Major Indian cities' energy-efficient homes sold an average of 24 days faster than conventional residences and paid a 5-12% premium according to a survey by the Indian Green Building Council.

The fact that many efficiency gains need little or nothing to execute makes them perhaps most interesting. With little investment needed, simple behaviour changes include thermostat adjustments, running appliances during off-peak hours, or better maintenance of current equipment can cut energy expenses by 10 to 15%.

Beyond Electricity: Comprehensive Energy Management

When we consider household energy, most people immediately think about electricity, that power running our appliances, lighting our homes, and charging our electronics. However, thorough energy management covers all kinds of energy passing through our houses and lifestyles, going well beyond just consumption.

Usually accounting for 14-18% of domestic energy consumption, water heating is a major yet sometimes disregarded energy cost. Independent of demand, traditional storage water heaters keep big hot water tanks running 24 hours a day. The U.S. Department of Energy estimates that switching to on-demand, tank-less water heaters can cut energy use by 24-34%.

Another significant portion of our overall energy footprint is transportation energy. In metropolitan India, the typical two-car household spends about ₹15,000-20,000 monthly on fuel. Without vehicle replacement, basic techniques such consolidating errands into single trips, keeping adequate tire pressure, and eliminating extraneous weight from vehicles can increase fuel economy by 5-10%.

Cooking energy–from gas, electricity, or other fuels – offers still more chances for economy. For many recipes, pressure cookers cut cooking time and energy by up to 70%. When covering pots when cooking can cut energy demands by 30 to 40%, matching pot sizes to burner sizes helps to prevent wasted heat.

Perhaps the most significant component of our overall energy footprint is embedded energy, that which is utilized to manufacture, move, and ultimately discard goods we buy. Making a smartphone requires more energy than operating it throughout its entire lifespan.

This embedded energy means that, more often than direct energy use, our choices about consumption affect our energy footprint. Purchasing durable things, fixing rather than replacing objects, and, if at all possible, selecting locally made goods help to lower this hidden energy usage.

Seasonal Strategies for Energy Optimization

Our energy management techniques should change with seasonal cycles, much as we modify our clothes and activities to fit changing seasons. Good energy efficiency calls for several strategies all year long, depending on living patterns, daylight hours, and temperature changes.

Summer energy control mostly addresses cooling efficiency. In warmer climes, cooling often accounts for 40–60% of summer energy use. Many passive techniques can greatly lower cooling needs outside of air conditioning levels.

The best way to stop summer heat gain is with outside shade. On windows facing west, window awnings can cut solar heat intake by up to 77%; on windows facing south, they can cut 65%. Temporary options like bamboo screens or shade sails give similar advantages without permanent change for homes without built-in shading.

Nighttime ventilation strategies take advantage of cooler evening temperatures. Opening windows on opposite sides of living areas produces cross-ventilation capable of flushing collected heat from the building mass. This natural cooling can keep comfortable temperatures with minimum mechanical help in areas with notable daily temperature variations.

Winter presents new energy concerns; heating usually represents the biggest energy cost. Often the best return on investment is from weatherizing windows, doors, and service entrances, sealing cracks around them. Small air leaks taken all together are like leaving a window half open all year long.

Strategic use of window coverings plays the opposite role in winter compared to summer. While covering windows after sunset helps maintain that warmth overnight, opening curtains and blinds during daylight hours on sun-facing windows catches free solar heat. Up to 25% of heat loss via windows can be prevented with sealed edge thermal curtains.

Maintenance done seasonally guarantees best system operation. Before summer, air conditioner maintenance and heating system maintenance before winter help to reduce efficiency losses usually resulting from neglect. Seasonally cleaning refrigerator coils, air filters, and ventilation systems helps appliances run as they should under different temperatures.

Energy Literacy: Knowing Your Consumption Habits

True energy smarts begin with energy literacy, your ability to read, comprehend, and react to your specific usage habits. Without this basic understanding, even the most sophisticated efficiency technologies or renewable systems can't function at their best.

Measurement is the key to energy literacy. Energy use precisely illustrates the adage of management: "you cannot improve what you do not measure." Standard utility bills merely give us monthly summaries, masking the very details that disclose the most promising possibilities for conservation.

Smart meters and energy monitors turn raw data into usable intelligence. They monitor usage in real-time, enabling you to link specific activities with their energy usage. Is operating the clothes dryer really a bigger electricity draw than the air conditioner? Is your water heater running more than it needs to? Real-time monitoring gives you definitive answers to these questions instead of making general assumptions.

The second phase of energy literacy is recognizing patterns. Examining when and where you consume energy identifies your distinctive consumption pattern. Maybe your home sees a rush of consumption around 6-8 p.m. as everyone comes home, electronics turn on, and meal preparation starts. Or your low-end consumption never goes below some threshold, signifying high phantom loads from electronics consuming power in standby mode.

Knowledge of these patterns makes targeted

instead of blanket efficiency tactics possible. Rather than trying in a general way to "use less energy," you can use your specific consumption patterns to highlight individual high-leverage opportunities. If your morning shower generates a daily hot water energy peak, a low-flow showerhead or heat-recovery drain may yield the highest payoff.

Building energy literacy, your skill at accurately observing, comprehending, and reacting to your own energy use patterns, lays the groundwork for all other energy optimizations. This foundational knowledge guarantees that your behaviour adjustments and efficiency investments address your real rather than inferred energy profile, optimizing both environmental and financial return.

Call to Action

This month apply three low-cost efficiency enhancements and do a home energy analysis.

Here's a useful application for energy intelligence in your own house as we wrap up this chapter:

Step 1: Perform Your Energy Audit

- Using a candle flame or incense stick, look for air leaks around windows, doors, and service doors – movement suggests air intrusion.

- Look at insulation levels in easily accessible places like crawl spaces or attics.
- Sort light bulbs, noting any last-minute halogen or incandescent varieties.
- Check refrigerator and freezer temperatures to guarantee best settings.
- Using an energy meter or by noticing gadgets with standby lights or heated power adapters when "off," find phantom loads.

Step 2: Identify Your High-Impact Opportunities

Using the results of your audit, compile a list of possible enhancements, concentrating on those that:

- Speak about your biggest energy consumption.
- Demand low financial input.
- One can start without certain knowledge or tools.
- Not going to greatly affect your comfort or convenience.

Step 3: Implement Three Improvements

Out of your list, three things you could finish this month. Typical high-return options consist in:

- Adding weatherstripping around windows and doors
- Changing your most often used lightbulbs to LEDs
- Including insulating around your water heater
- Setting up smart power strips for home offices and entertainment areas
- Mastic or metal-backed tape can be used to seal ducting
- Setting low-flow showerheads

Which three energy-saving ideas will you start this month?

Chapter 5

Quieter Existence: Addressing Sound Pollution

The Unseen Health Impacts of Noise Pollution

Imagine attempting to focus on a difficult project while a jackhammer pounds outside your window. Alternatively trying to sleep while traffic hums past, broken sometimes by car horns. These situations are more than just irritations; they are a major environmental risk identified by the World Health Organization as the second most harmful environmental stressor influencing human health, following only air pollution.

The chart below illustrates common urban noise sources and their potential health impacts:

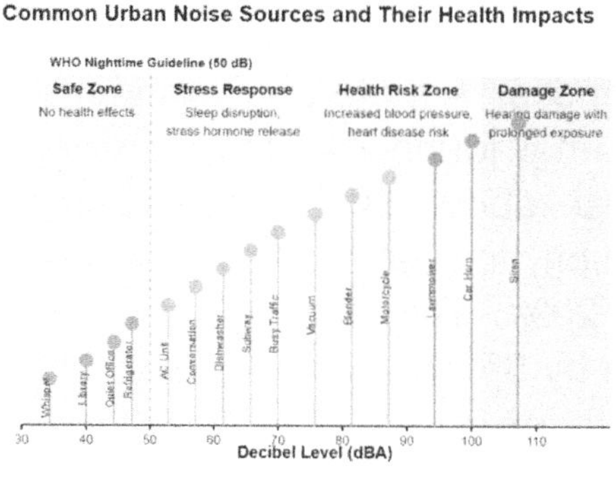

Source: World Health Organization (WHO),Environmental Noise Guidelines for the European Region. Chart adapted from WHO data on urban noise exposure thresholds and associated health outcomes.

Unlike obvious toxins that we can see building up in our surroundings, noise pollution is mostly undetected yet its consequences are profoundly felt in our bodies and brains. Science increasingly shows that, even in sleep, our auditory processes never really "switch off." Our ears are still absorbing and interpreting sounds, which sets up physiological reactions capable of upsetting important body systems.

The health effects start with reactions to stress. Our bodies release cortisol and adrenaline, the same "fight or flight", hormones set off by loud or unexpected noises. While this reaction is temporary, long-term noise exposure raises these stress hormones, which fuels a chain reaction of health issues.

Among the most critical are cardiovascular consequences. A meta-analysis in the European Heart Journal found that long-term exposure to traffic noise above 50 decibels, akin to a quiet conversation, increases the risk of heart attack by about 20%. Risk for coronary heart disease increases by 8% for every 10-decibel increase in road traffic noise.

Another important effect is disturbance of sleep. Noise induces changes from deeper to lighter sleep stages even if it does not completely wake us. According to a study by the German Aerospace Centre, nighttime aircraft noise particularly interfered with REM sleep and slow-wave sleep, the periods most important for physical restoration and memory consolidation.

Children especially find the repercussions to be quite worrisome. A landmark study tracking schools near airports found that students exposed to higher aircraft noise showed significant impairments in reading comprehension, memory, and problem-solving abilities. These cognitive effects continued even with socioeconomic considerations taken into account.

Our inclination to think we've "gotten used to it" makes noise pollution especially sneaky, but physiological studies show otherwise. Whether or not we are aware of it, our bodies still mount stress reactions to noise even after years of exposure. The body merely accumulates damage; it never really adapts.

Urban Design for Sound Management

When we consider city design, visual components usually take front stage – the height of buildings, the width of streets, the existence of parks. Every urban design choice, however, also influences how sound travels through the surroundings and generates acoustic effects that impact millions of daily occupants.

Rather than deliberate design, modern urban soundscapes sometimes arise from design mistakes. Hard, reflecting surfaces like glass, concrete, and asphalt produce canyons where sound echoes rather than absorbs. Straight, wide streets let noise travel unhindered. Mixed-use zoning places homes right next to businesses with opposing acoustic needs.

However, a growing movement of acoustic-conscious urban design is demonstrating that cities can be both vibrant and sonically comfortable. These methods seek for balanced soundscapes where necessary urban activity can coexist with human welfare rather than for artificial silence.

Among the most versatile and successful urban sound techniques are vegetative barriers. By 5-10 decibels, dense plantings of trees and bushes along roadways can help to lower perceived traffic noise, a change that would feel like halving the noise level to human hearing. Soft leaf surfaces absorb sound energy, complex branch structures deflect sound waves, and the visual screening produces a psychological barrier that lowers perceived noise levels. These green buffers operate through several channels.

The form of the urban surroundings itself greatly affects the sound propagation. Instead of continuous rows, buildings set in a checkerboard pattern generate acoustic "shadow zones" where sound levels drop. Angled or curved streets stop noise from unhindered straight-line travel. Different building heights produce a more complicated acoustic environment that stops standing sound waves from developing between parallel facades.

Urban acoustics rely critically on surface materials. While it improves water drainage, porous asphalt can cut tire noise by 3-5 dB over conventional pavements. Made from perforated materials or vegetation, sound-absorbing facades help to stop sound reflection between buildings.

This is taken further in the idea of "sound scaping," which purposefully adds uplifting sounds to cover inevitable noise. While providing visual appeal, the sound of running water from ornamental fountains efficiently covers traffic noise. Natural sound markers created by spaces meant to promote songbirds most people find pleasing instead of annoying.

Transportation Noise: Solutions and Alternatives

The tires' steady hum on asphalt. The sound of planes roaring above. The loud horns amid rush hour traffic. For most urban and suburban people, transportation noise rules the acoustic scene of modern life and is the most common cause of noise pollution.

In most cities, road traffic alone causes more than eighty percent of the noise pollution. With a loud footprint spanning hundreds of meters in all directions, one roadway can impact thousands of houses along its course. Communities distant from airports themselves are affected by aircraft noise; flight paths create corridors of disturbance.

The problem with traffic noise is essentially design one. With noise consequences seen as inevitable side effects, most cars were built with speed, convenience, and cost as main factors of importance. New technology and methods, however, are showing that quiet transportation systems can also be efficient.

Perhaps the most important transportation noise revolution is that of electric cars (EVs). EVs almost completely eliminate engine noise at speeds less than 30 km/h, generating about 50% less noise than equivalent internal combustion vehicles. In metropolitan settings with many stops and starts, this benefit is especially important.

Many technical methods provide significant noise reductions for conventional automobiles. Low-noise pavement surfaces cut tire noise by 3-7 dB utilizing rubber-modified asphalt or open-graded friction courses. By surrounding power units with sound-absorbing materials, engine encapsulating technologies can cut engine noise by up to 10 decibels.

At the community level, noise from transportation demands for coordinated responses. Copenhagen's all-encompassing mobility plan calls for public transit for medium distances, bicycle and walking for short excursions, and auto reserves for trips when other choices are not possible. By moving to quieter alternatives for many everyday trips, this modal hierarchy organically lowers traffic noise.

Most crucially, maybe, traffic noise solutions complement other sustainability objectives. Policies that lower car reliance promote public health by means of more physical activity combined with noise reduction, therefore improving air quality and lowering carbon emissions.

Cultural Aspects of Noise Tolerance

Why might noise levels deemed unacceptable in Tokyo seem to be very reasonable in Cairo? Why do some towns accept street vendor sounds while others forbid them totally? The solutions reside not only in physics but also in the intricate interaction of society, history, and culture that shapes our interaction with sound.

Noise tolerance shows amazing range depending on culture and geography. Self-reported noise irritation differed greatly even in cases when recorded sound levels were comparable, according to a comparative study spanning 36 countries. While participants from Latin American countries generally reported lower discomfort at similar noise levels, Japanese respondents reported great annoyance at noise levels that Dutch participants found just mildly disruptive.

These variations reveal strongly ingrained cultural ideals about public behaviour, personal space, and community ties. Strongly emphasizing group harmony cultures often show more sensitivity to noise as possible disturbance. On the other hand, societies honouring active public life could view some noises not as noise but rather as the natural acoustic tapestry of community vitality.

Over decades, historical events help to define these impressions. Long histories of dense urban living often translate into greater tolerance for proximity sounds–conversations across apartment walls, footsteps on staircases, children playing in common areas. These sounds start to be seen as ordinary features of community life instead of invasions.

Even what defines "noise" against "sound" differs culturally. Bangkok's street markets, with merchants calling their goods and negotiating at conversational decibels that overlap and overlay, generate soundscapes that residents often find suitable and even vital to the experience. One could argue that the same auditory environment taken to a Scandinavian metropolis is unacceptably chaotic.

Noise policy and management are largely affected by these cultural differences. Strategies that succeed in one cultural setting could not be at all applicable in another. Noise rules created in Northern Europe cannot be imported to Mediterranean or Southeast Asian towns without modification to fit local cultural settings.

Good strategies acknowledge these cultural aspects instead of considering noise as a technical issue only. Where top-down technological requirements fall short, community-based noise management including locals in establishing acceptable soundscapes usually finds success.

Case Study: Cities Transforming Their Soundscapes

Beyond theoretical debates, some innovative cities have used all-encompassing noise control strategies to show how focused action can change urban soundscapes. These case studies provide communities looking for quieter, better surroundings both inspiration and doable plans.

Among urban acoustic transformations, Zurich, Switzerland appears to be the most ambitious one. Over most of the city, noise levels exceeded health recommendations; Zurich launched a multi-decade soundscape restoration effort combining public involvement, infrastructure, and legislation.

Reconfiguration of transportation was central to

Zurich's approach. Currently representing 91% of the overall road network, the city methodically lowered speed restrictions to 30km/h throughout 400 kilometers of residential streets. By 2-4 dB, this one adjustment lowered noise level for 50,000 people without appreciably changing travel times. Low-noise asphalt during routine road maintenance can be installed to progressively produce a quieter road network free from dedicated resurfacing operations.

Zurich implemented both physical infrastructure improvements and strict vehicle noise regulations, featuring automated sound monitoring systems that detect and locate excessively noisy vehicles. Nowadays, the planning department of the city mandates acoustic impact studies for any significant development project together with strict noise limitations that have to be satisfied before building authorization.

The outcomes are really outstanding. From 22% to 14% the proportion of Zurich inhabitants exposed to too much evening noise dropped between 2010 and 2020. Previously thought of being acoustic "sacrifice zones" next to important infrastructure, these areas have become viable communities because to deliberate interventions.

With its creative superblock project, Barcelona, Spain has tried a different but equally successful method. The superblock form generates acoustic oasis inside the dense urban fabric by limiting vehicle access inside neighbourhood blocks and reclaiming street space for community use.

Noise levels have dropped between 5 and 15 dB within Barcelona's superblocks, therefore changing the acoustic environment from one dominated by traffic to one where human voices, birds, and natural noises become audible. These better soundscapes have supported more general social advantages like more older people using public areas, more children playing outside, and greater community bonds created by natural contact.

Under its "Sounding Brighton" project, Brighton and Hove in the United Kingdom have pioneered community-based soundscape design at a smaller scale. This method aggressively finds and protects good acoustic elements – the sounds of seagulls along the shore, street performers in suitable places, and the soft background hum of cafe conversations – rather than concentrating just on lowering undesirable sounds.

Through great community involvement, neighbors found which sounds enhanced their experience of the city and which degraded it. This subtle approach produced context-sensitive adjustments, moving delivery traffic away from residential areas during early morning hours and so maintaining the vibrant sounds of the beach-side promenade during daytime.

Personal Practices for Reducing Noise Production

Although infrastructure and regulatory changes offer the basis for calmer communities, our individual actions help to define the acoustic environment we share. Every one of us makes hundreds of everyday decisions about sound that either add to noise pollution or help to preserve acoustic quality for others and ourselves.

Our most important personal noise influence is that of transportation decisions. Driving styles significantly affect the production of vehicle noise; fast acceleration produces up to 15 dB more noise than slow speed changes. Just avoiding abrupt acceleration and deceleration helps to simultaneously lower the noise signature and fuel consumption of your car.

Horn use is a particularly focused kind of noise pollution. A single horn blast can reach 100-110 decibels, loud enough to inflict hearing damage with sustained exposure and most definitely enough to disturb sleep or concentration over great distances. One of the easiest but most effective personal noise reduction techniques is limiting horn use to real safety events instead of displays of frustration.

Beyond mobility, choices about home appliances have great impact. Noise ratings on modern appliances let users choose quieter models.

Usually running three to seven decibels quieter than earlier models, energy-efficient air conditioners. While delivering full power when needed, variable-speed fans and pumps preserve reduced noise levels during typical operation.

The timing of loud activities offers still another opportunity for deliberate personal improvement. Setting food preparation, washing cycles, and vacuum cleaning for noon instead of early morning or late evening helps neighbours' sleep and concentration span. In residential environments, being aware of footfall sounds during nocturnal hours, using soft footwear or changing walking patterns, can significantly enhance the quality of sleep of neighbors.

Technology for communication presents opportunities as well as problems. Shared acoustic rooms are created in part via speaker phones, video calls without headphones, and media running through device speakers. Maintaining acoustic politeness requires using headphones for personal media use, locating private areas for phone talks, and lowering gadget alert volumes to the minimum required level.

These personal habits simply involve awareness of when, when, and how we carry out regular daily tasks; they do not call for giving up activities or tolerating difficulty. Without tight rules or major lifestyle changes, these little behavioral modifications taken together can greatly enhance the acoustic environment.

Soundproofing Techniques for Urban Living

Urban settings will always feature certain sounds outside of our reach even with the best community noise control. Soundproofing methods offer a last layer of protection, thereby establishing acoustic sanctuaries inside our houses independent of outside environment. From basic, low-cost treatments to more all-encompassing solutions for demanding surroundings, these methods span a spectrum.

The following graph presents some house soundproofing methods together with their efficiency:

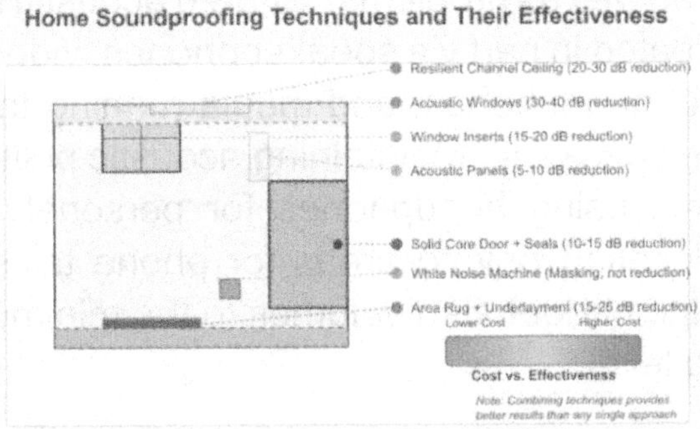

Source: National Research Council Canada (NRC), Building Acoustics: Noise Reduction Techniques for Residential Buildings (2020). Chart adapted from NRC research data on

sound transmission reduction in residential environments.

Usually, windows show the weakest acoustic location in any kind of building shell. While even ordinary double-glazing provides only modest improvement, single-pane windows allow up to 90% of outside noise to be transmitted. But by 30 to 40 decibels, acoustic windows with different glass thicknesses, more air gaps, and specialist seals can help to lower noise transmission, thereby turning even a roadside apartment into a rather quiet environment.

Window inserts offer a temporary fix free of permanent changes for renters or individuals with limited funds. By mounting these clear acrylic panels inside current window frames, an extra air barrier is created that, at a fraction of replacement window prices, can cut noise transmission by 50-70%.

Another typical acoustic flaw is seen in doors. Unbelievably large levels of noise can be transmitted even from little spaces around door edges. Although they cost under $20, door sweeps that close the gap between door bottom and floor usually cut noise transmission by five to seven decibels. While solid-core doors offer far better sound isolation than hollow-core options, perimeter weatherstripping similarly covers gaps around the sides and top of doors.

Wall treatments have acoustic as well as ornamental value. Bookcases loaded with books absorb significant sound energy and provide useful storage capacity. Particularly those with pleats or rough surfaces, fabric wall hangings absorb very little sound. Acoustic panels created from recyclable materials or compressed fiberglass can be used as decorative accents to significantly lower room acoustics.

Technology provides still another choice for achieving acoustic comfort. Less obtrusive white noise devices produce steady, broadband sound that conceals irregular noises like traffic or conversations. Using microphones and speakers, active noise cancellation systems can lower continuous low-frequency sounds like traffic rumble or HVAC system noise that conventional soundproofing finds difficult to handle.

Even people living in crowded cities can design acoustic sanctuaries supporting rest, focus, and health by combining these several approaches depending on their needs and limits. These enhancements reflect not luxury but rather required infrastructure for both physical and mental health in noisy surroundings.

Call to Action

Start a “Necessary Horns Only” campaign in your neighbourhood featuring obvious reminders in busy traffic zones.

As we conclude this chapter, here's a practical way to address one of the most unnecessary yet pervasive forms of noise pollution:

Step 1: Assess the Situation

- Use observation to find high-horn-use places in your neighbourhood.
- Note patterns: Is horn use mostly during rush hour? Not far from some definite crossings? Connected to certain driver actions?
- Think about whether needless honking results from local circumstances, greeting customs, or frustration.

Step 2: Design Your Campaign

- Make basic, graphic signs with unambiguous messaging like "Honk Only for Safety" or "Horn = Emergency Only."
- Create educational materials on how horn sounds affects quality of life and health.
- Get ready with a quick presentation for local leaders or community meetings.

Step 3: Build Community Support

- Approach community wellness programs, school parent groups, or neighbourhood associations.

- Engage local companies in impacted areas who might be in favour of more subdued streetscapes.
- Choose a few passionate volunteers to assist with placement and upkeep of signage.

Step 4: Implement Visual Reminders

- With appropriate rights, position signs at key points such as intersection approaches.
- Think about temporary chalk stencils on sidewalks next to areas of high honking.
- Share window signs for homes and businesses in impacted hallways.

With just basic awareness and behaviour modification, what unneeded noise in your neighbourhood could be eliminated?

Chapter 6

The Microplastic Problem

Concealed Sources of Microplastics in Everyday Products

They're in your morning routine, your washing, even your cup of tea. Microplastics, plastic pieces less than 5mm in size, have invaded almost every part of contemporary life, frequently concealed in products we'd never imagine.

Think about your morning face wash. Many exfoliating scrubs contain microbeads, tiny plastic spheres designed to slough off dead skin. Though some countries have banned them, countless products still include these persistent pollutants. Your toothpaste might harbour them too; those coloured specks marketed as "whitening crystals" are frequently plastic.

Synthetic clothing is another significant source. With every wash of polyester, nylon, or acrylic clothing, thousands of microfibers release into wastewater. Up to 700,000 synthetic fibres can be released from a single load of laundry. Synthetic jackets were found by the Environmental Science and Technology journal to release on average 1.7

grams of microfibers with each wash, nearly as heavy as a paper clip.

Even tea bags have a role. High-end brands of tea make use of "silken" pyramid bags, which are made not of silk but of nylon or PET plastic. Steeping them in boiling water, the bag in every cup can deliver billions of microplastic pieces into your brew, McGill University research has shown.

Paint is a less apparent offender. As its weathers on the outside of houses, it degrades into microplastic particles that are washed down storm drains. The same occurs with road markings and boat paints. The European Chemicals Agency estimates that 42% of primary microplastics emitted into the environment are due to abrasion of plastic products in use, tire wear, which deposits microplastic-contaminated dust along all roadways.

"Most unsettling is that all these sources encompass us every moment, leaving a hidden plastic mark much greater than the apparent plastic items we actively use and reject," says environmental scientist Dr. Sherri Mason.

The Lifecycle of Microplastics: From Sink to Ocean

The lifecycle of microplastics in the world around us takes many roads, all ending up at one place: our waterways, and eventually the ocean.

The trip often starts at your washing machine or sink. Traditional wastewater treatment facilities were not meant to screen particles as tiny as microplastics. Research conducted by Water Research revealed that despite treatment trapping 90-99% of microplastics, the high amounts treated result in billions of microplastic particles still entering waterways each day. One treatment facility can have millions of microplastic particles enter receiving waters daily.

Rain carries microplastics off roads, buildings, and landfill sites into storm drains. These tend to flow straight into rivers without treatment. The microplastics move downstream, depositing in riverbeds or flowing on to the coast.

When microplastics reach the ocean, they don't vanish. Ocean currents distribute them worldwide, scientists have discovered microplastics in Arctic ice and on the bottom of the Mariana Trench, the lowest point in the oceans. Winds can transport lighter particles thousands of kilometres from where they started.

Along the way, microplastics both increase in number and in toxicity. "In water, plastics don't biodegrade, they photodegrade, breaking into ever-smaller pieces while retaining their polymer structure," says Dr. Chelsea Rochman of the University of Toronto. These small particles take up persistent organic pollutants, basically becoming poisonous pills that marine animals confuse with food.

This spread has resulted in microplastics becoming fully ubiquitous in the environment. They've been discovered in drinking water from New York to Mumbai, in Antarctic snow, and even in the placenta of unborn children. The cycle has completed itself, and now humans are consuming the very microplastics we've discharged.

Alternative Products and Materials

Confronting the microplastic challenge requires replacing problematic materials with better alternatives, many of which already exist and simply need wider adoption.

Natural fibers are a valuable alternative to synthetic fabrics. Organic cotton, hemp, and wool biodegrade naturally if they find their way into the environment. Think about fashion, but also function and end-of-life consequence when selecting fabrics. A linen blouse may be more expensive upfront but will generally outlast synthetic ones without shedding microplastics in the wash.

For body care products, natural exfoliants such as ground apricot kernels, walnut shells, or even plain sugar make great plastic microbead substitutes. Some companies, including Lush and Aesop, have already led the way with fully plastic-free formulations and shown commercial viability for these substitutes.

Compostable polymers and plant-based plastics are promising alternatives to traditional plastics. Composed of materials such as corn starch, sugar cane, or crop waste, many can be broken down in industrial composting. But careful evaluation is required, not all bioplastics break down in home composting or in the ocean.

For household cleaning, natural materials tend to be more effective than plastic-based products. Loofah sponges, coconut fibre scrubbers, and cellulose cleaning cloths eliminate the use of synthetic scouring pads that release microplastics. "Our ancestors had clean homes without plastic for centuries," says sustainable living activist Maya Gordon. "We're just rediscovering those techniques with some modern enhancements."

Advocacy and Consumer Pressure for Change

Individual choices matter, but system-wide changes require collective action. Advocacy and consumer pressure have proven remarkably effective in addressing microplastic pollution when properly coordinated.

The effective campaign to ban microbeads illustrates this strength. While researchers initially reported microbead pollution in the Great

Lakes in 2012, public awareness was low. As environmental organizations, scientists, and active citizens joined efforts, public pressure built rapidly. Major companies such as Johnson & Johnson and Unilever made voluntary phase-outs, followed by legislations to ban them in many countries by 2018.

Product labelling is a key advocacy effort as well. Today, tracking products with microplastics calls for technical proficiency in ingredient nomenclature. Clear labelling efforts on plastic content allow consumers to make educated buying decisions and generate market pressure.

"Consumer pressure has proven to be one of the most effective drivers of corporate environmental change. When consumers demonstrate clear preferences through purchasing decisions, companies respond with remarkable speed to maintain market share" (Lyon & Maxwell, 2011, *Corporate Environmentalism and Public Policy*, Cambridge University Press).

Digital technology has amplified consumer voices. Software such as Beat the Microbead enables shoppers to scan product barcodes to test for microplastic content, while social media campaigns can quickly spread the word about troublesome products.

When communicating with companies, targeted feedback is most effective. Instead of open-ended complaints, messages stating precisely which products you've stopped buying, and why, create clear market signals that companies get.

Case Study: Communities Successfully Reducing Microplastic Pollution

While the microplastic issue seems daunting, communities all over the world have implemented effective reduction strategies, proving that grassroots action can truly make a difference.

The Vamizi Island coastal village of Mozambique was severely affected by microplastic pollution that inflicted damage on marine life and tourism. Rather than waiting for national legislation, local authorities implemented a threefold approach: education in schools, a beach plastic collection incentive program, and cooperation with hotels to prohibit single-use plastics. Beach plastic pollution was reduced by 73% within two years, and microplastic levels in waters around the coast were measurably reduced.

In Bengaluru, India, the Jakkur Lake restoration project shows the contribution of urban wastewater management towards microplastic reduction. The people implemented a constructed wetland system that naturally purifies water before flowing into the lake. This natural solution removes up to 97% of microplastics from the inflowing wastewater while improving habitat restoration and beautification.

The German town of Freiburg wanted to avoid tire microplastics, arguably one of the most common sorts that can be found in urban environments. By increasing car-free areas, enhancing public transport, and using green infrastructure to capture road runoff, the community minimized tire-related microplastic contamination by an estimated 42% within five years.

"The most effective community projects blend education with infrastructure enhancements and policy shifts," says environmental planner Ruth Patel. "They make the sustainable option the convenient option."

Microplastics in Food and Water: Unraveling the Health Impacts

We now consume microplastics regularly in food, in water, and even in the air we breathe. "According to a 2019 World Wildlife Fund report conducted by the University of Newcastle, a typical human consumes approximately 5 grams of plastic per week - roughly equivalent to a credit card. This analysis, titled 'No Plastic in Nature: Assessing Plastic Ingestion from Nature to People,' synthesized data from multiple studies on microplastic consumption. "Among the common foods that have high rates of microplastic contamination are seafood, bottled water, beer, and salt. It was shown by a study that people who regularly consume shellfish can invest up to 11,000 microplastic particles annually. Orb Media's 2018 global study "Invisibles: The Plastic Inside Us," which tested 259 bottles from 11 brands across 9 countries and found bottled water contained an average of 22 times more microplastic particles than tap water.

Their health impact is not yet comprehensively known, but the scary evidence keeps growing. In laboratory experiments, microplastics have been shown to be able to cross cell membranes, and they are able to cause inflammation, immune response, and oxidative stress. Microplastics are also capable of serving as a vector for other pollutants like heavy metals and persistent organic pollutants, which adhere to the surface of the plastic.

"The most concerning aspect of microplastics is that they stick around in body tissues," says toxicologist Dr. Sarah Janssen. "While most other types of pollutants, which the body can metabolize and eliminate, microplastics can amass in organs such as the liver, kidneys, and even brain tissue."

Most concerning is the possible developmental effect on children. Animal research indicates early exposure can interfere with hormone systems and brain development. Although human research is only just starting, caution appears wise in light of widespread exposure.

DIY Home Solutions for Filtering and Minimizing Microplastic Exposure

While systemic changes move forward, home-based practical solutions can greatly minimize your individual microplastic exposure and contribution.

- For washing clothes, a significant source of microplastics, filtering options vary from basic to high-tech. For washing clothes, a significant source of microplastics, filtering options vary from basic to high-tech. Microfiber-catching washing bags trap 90% of microfibers from synthetic garments while washing.

- Drinking water is another point for intervention. Standard water filters don't specifically aim at microplastics, but reverse osmosis systems strip away particles as small as 0.001 microns, adequate for most microplastics. Activated carbon filters cut microplastic content, though less completely, at a more affordable cost.

- Basic behavioural modifications supplement filtering measures. Reducing washing of synthetic clothing and lowering the temperature lessens microfiber shedding by 40%. Opting for natural fibre garments on new purchases increasingly minimizes microplastic sources in your closet.

- For food preparation, minimizing plastic containers, especially with hot foods, minimizes microplastic contact. Glass, stainless steel, and ceramic options remove this contamination source while sometimes lasting longer than plastic equivalents.
- Even driving behaviour influences micro-plastic pollution. Gentle acceleration and braking minimize tire wear, lowering the microplastics shed onto roadways and flushed into waterways. "The same driving methods that conserve fuel also minimize tire microplastic production," says transportation researcher Alex Patel.

Call to Action

Inspect your bathroom and kitchen products for microplastic content and substitute the three worst offenders with sustainable alternatives.

As we finish this chapter, here's a useful way to cut your microplastic footprint:

Step 1: Do Your Audit

- Collect all personal care products such as soaps, scrubs, toothpastes, and cosmetics.
- Scan ingredients for telltale microplastic ingredients: Polyethylene (PE), Polypropylene (PP), Polyethylene terephthalate

(PET), Polymethyl methacrylate (PMMA), or Nylon.

- Check kitchen cleaning products, particularly scouring pads and synthetic sponges.
- Record all products with microplastics, prioritizing by use frequency.

Step 2: Investigate Alternatives

For each product containing microplastics, identify at least two natural alternatives.

- Consider effectiveness, cost, and availability when assessing substitutions.
- Search for products with explicit "microplastic-free" or "plastic-free" labelling.

Step 3: Replace Highest-Impact Items

- Begin with your three most used products with microplastics.
- Substitute with selected alternatives, preferably one by one to acclimatize to any variation.
- Monitor your experience with each substitution to optimize subsequent decisions.

What three microplastic-containing products will you substitute this month?

Chapter 7

The Convenience Economy: Service and Sustainability in Balance

The Cost of Convenience: Economic and Ecological

We've grown so used to unparalleled convenience. With a swipe or two on our phones, we can call forth food, produce, household supplies, and nearly every service to our front door. This revolution of convenience has revolutionized our everyday lives, but it has unseen costs that reach well beyond the delivery charges and gratuities we pay.

The convenience premium is high financially. Food ordering apps usually add 40-50% to the cost of a meal through aggregated service fees, delivery fees, and price inflation on the menu. The Financial Times analysis discovered that similar grocery products were 23% more expensive, on average, through delivery apps compared to retail stores. The costs add up quietly, the average urban Indian family paying ₹5,000-7,000 a month on convenience services that they could easily do themselves.

The impact on the environment is also disconcerting. Last-mile delivery is responsible for a considerable part of city traffic congestion and carbon footprint. In a 2021 World Economic Forum report, it was estimated that without a solution, urban areas will have 36% more delivery vehicles by 2030, increasing carbon emissions by 32%.

Convenience service packaging waste has come to staggering levels. In the United States alone, food delivery creates an estimated 282 million kilograms of plastic waste each year as per the Environmental Protection Agency. In India, Delhi and Bengaluru alone collect tens of thousands of food delivery containers every day, with the majority going to landfills despite claims of recyclability.

"We've made systems where it's more convenient to dispose of than reuse," says sustainable design specialist Aamir Sharma. "The real cost involves not only production and disposal, but also the harvesting of virgin resources and the ongoing pollution that we leave behind in our environment for centuries."

Redesigning Food Delivery Systems

Food delivery is one of the most conspicuous manifestations of convenience culture, and one of the most fertile opportunities for environmental redesign. There are already a number of promising new models under development that preserve convenience while slashing waste.

Reusable container systems are gaining popularity in urban centers. Firms such as Tiffin in Mumbai and DabbaDrop in London employ stainless steel containers, which are picked up, cleaned, and circulated. The system is based on India's historic dabbawalla system but adds modern logistics and digital monitoring.

These systems form closed loops that cancel out single-use waste. Based on lifecycle analyses by the Journal of Cleaner Production, a stainless-steel container must only be reused 20-30 times to surpass single-use options in all environmental measures, and most are constructed for hundreds of uses.

Consolidation delivery platforms are another efficiency innovation. Instead of single restaurants sending out individual delivery cars, centralized "ghost kitchens" house a number of restaurant brands under one roof and have a single driver deliver orders from various restaurants at the same time. This setup trims delivery car emissions by as much as 30% based on logistics analytics company Urbantz.

Technology now facilitates new opportunities for optimizing last-mile delivery. In high-density city centres, bicycle and e-bike courier services like Zomato's emissions-free vehicles in Delhi provide quick delivery with no direct emissions. Delivery robots are now being tested in a number of cities for short-distance deliveries that can potentially wipe out vehicle trips altogether for orders that are nearby.

Consumer Choice as Market Driver

While regulations and business have roles in establishing sustainable systems, customer decisions ultimately establish the direction for the market. What we purchase conveys influential messages about our values.

When sufficient consumers express preference for sustainable alternatives, markets react swiftly. Following consumer campaign feedback on excessive packaging, some of the large food delivery companies launched opt-out features for cutlery and condiments. Zomato reported that in three months since launching this feature, 80% of customers opted to forego unnecessary items, averting millions of plastic utensils from entering the waste stream.

Rating systems are another highly effective feedback mechanism. When delivery platforms included restaurants' sustainability ratings, showing those with environmentally friendly packaging or plastic-free meals, customer choice patterns changed significantly. Restaurants with better sustainability ratings experienced order count growth of 5-11% in the first quarter following implementation, as reported by industry data.

"Customers tend to think their own decisions won't make a difference, but the collective preferences of customers create huge market changes," says behavioural economist Dr. Nisha Rao. "When companies realize sustainable alternatives become competitive strengths instead of expense items, innovation happens very fast."

Digital Solutions for Physical Waste

The same digital revolution that facilitated the convenience economy also provides solutions to offset its environmental cost. Technology presents many opportunities to separate convenience from waste.

Paperless and digital receipts cut down on huge paper waste. A typical restaurant meal produces 0.5-1.5 grams of paper in receipts alone. With billions of delivery orders around the world, digital documentation keeps thousands of tons of thermal paper waste from occurring, especially pertinent as most thermal paper contains BPA or BPS chemicals that taint recycling streams.

Predictive analytics assist in right-sizing production and minimizing food waste. Delivery platforms now employ artificial intelligence to forecast order volumes, enabling restaurants to prepare the right quantities instead of overproducing to prevent stockouts. Swiggy's AI-driven demand forecasting system assisted partner restaurants in minimizing food waste by an estimated 11% in its first year of use.

Intelligent packaging with digital stamps facilitates improved waste management. QR codes and other digital data in packaging can include detailed recycling guides tailored to local facilities, leading to higher rates of proper disposal. In Bengaluru pilot projects, digitally-marked food containers had 23% higher rates of proper recycling compared to unmarked counterparts.

Blockchain-based tracking of supply chains introduces unprecedented transparency into consumption habits. Businesses such as Provenance enable consumers to authenticate claims of sustainability by viewing assured supply chain information for products. Applied to seafood delivery in Chennai, the technology improved choice of sustainably-sourced products by 18%.

Case Study: Businesses Thriving on Sustainable Convenience Models

Opposed to popular lore that sustainability calls for sacrifice, visionary companies are demonstrating that environmental stewardship and convenience are mutually reinforcing and not opposing values.

Dabbawallas of Mumbai represent perhaps the oldest sustainable delivery model, operating successfully for over 125 years. Each day, this network delivers more than 200,000 home-cooked meals in reusable containers throughout the city using only bicycles and trains. The system achieves remarkable efficiency, only one error per 8 million deliveries, while remaining emissions-free and waste-free. Their model has inspired numerous modern adaptations globally.

In Singapore, barePack created a tech-enabled reusable container system that partners with over 150 restaurants. Customers use an app to borrow containers when ordering takeaway, returning them to any participating location. In just two years of operation, the system has prevented more than 500,000 single-use containers from entering the waste stream while maintaining complete convenience for consumers.

Loop, a worldwide program created by Terracycle, teams up with big brands to supply products in high-quality reusable containers. They are retrieved, sanitized, and refilled in a contemporary take on the original milkman idea. Early consumer testing revealed 60% of users kept using the service beyond the pilot duration, attributing it to the quality containers being on par or better than single-use competition.

“What all these successful models have in common is that they are interested in building on, not just replacing, the convenience experience,” says sustainable business analyst Maya

Chopra. "They understand that in order to scale, sustainable options need to be as good or better than the current state."

Packaging Innovation: What's Coming and How to Support It

The packaging industry is undergoing rapid transformation, with innovations addressing both materials and systems. These developments promise to maintain convenience while dramatically reducing environmental impact.

Compostable bioplastics made from farm waste are a promising horizon. Compostable bioplastics made from farm waste are a promising horizon. Some innovators are producing food packaging made from rice and wheat bran that would otherwise be wasted. Unlike most first-generation bioplastics that needed industrial composting units, these new materials degrade in home composting systems or in nature within 60-90 days.

Edible packaging is a new strategy to end waste completely. A study in the Journal of Food Engineering has shown that protein-based packaging films such as casein and whey, and polysaccharide-based films like starch and chitosan, can form edible barriers for food items. The materials proved to prolong shelf life without compromising on safety to eat. Moreover, seaweed materials are also being researched for their application as edible and biodegradable packaging options that would take weeks to decompose in seawater, according to some prototypes.

Reusable packaging systems are expanding through standardization. The Foundation for Reusable Systems has created open-source standards for the material and size of containers so that several service providers can use compatible containers. Compatibility generates network effects making reuse more economic and convenient across geography.

"Supporting these innovations requires more than passive consumer interest," explains packaging engineer Vikram Seth. "Early adopters create the demand signals that allow sustainable packaging to achieve economies of scale."

The Psychology of Convenience and How to Rewire It

Our relationship with convenience is deeply psychological. Understanding these mental patterns helps us make more conscious choices that align convenience with sustainability rather than placing them in opposition.

The human brain is programmed to be energy-conservative, so convenience is naturally enticing. Behavioural science demonstrates that we overestimate up-front benefits (convenience, time savings) and discount costs that are ahead in time (environmental destruction, resource usage). This is a temporal error that makes convenience especially hard to resist, even when we rationalize its price.

Neuroplasticity research delivers good news: we are able to rewire these automatic proclivities by persistent training. As we repetitively make sustainable selections together with associated rewarding experiences, alternative neural tracks firm up and continuously alter what constitutes convenience and pleasantness.

"Convenience is subjective and not objective," claims environmental psychologist Dr. Anand Patel. "What comes naturally is most controlled by what our habits are as well as how we believe, not fixed amounts of work."

We can tap into this subjectivity by recasting sustainable decisions. Instead of perceiving reusable containers as a hassle to remember, we may find them an affirmation of personal values. Instead of perceiving plastic-free alternatives as constraints, we may find them as premium experiences, similar to how we've been culturally conditioned to perceive organic food.

Call to Action

Begin to specify precisely what disposable items you require (if any) when placing takeout or delivery orders, and give companies feedback on their default behaviours.

As we end this chapter, here's an easy but impactful step to cut convenience waste:

Step 1: Audit Your Takeaway Orders

- For your next three delivery or takeaway orders, record all the disposable items that come with them.
- Mark down which ones you actually used and which went directly to waste.
- Pay attention to duplicate packaging (bag inside a bag) and auto-added items (napkins, condiments).

Step 2: Tailor Future Orders

- Utilize the "special instructions" space to state exactly what you desire.
- Ask them to leave no plastic utensils if you'll be eating in-house.
- Say no to added napkins, packets of sauce, or other throwaways that you don't want.
- On repeated orders, save these choices as your profile settings.

Step 3: Leave Direct Feedback

- Use the feedback or rating mechanism to

rate packaging habits.

- Praise companies that provide eco-friendly alternatives or respect your requests.
- Suggest better options to companies that continue to provide unnecessary disposables.
- Spread the word among your friends to make personalized ordering the norm.

What disposable items will you cut out of your next delivery order?

Chapter 8

Financial Sustainability: The Personal-Planetary Connection

The Economics of Overconsumption

Our economy has been founded upon a disturbing paradox: infinite growth on a finite planet. The conventional economic formulas we've operated under for centuries have assumed natural resources as limitless and the ability of the environment to accept waste as limitless. This underlying disconnect is responsible for many of our sustainability issues today.

At a personal level, this takes the form of over-consumption, purchasing more than we require or can use effectively. The statistics are alarming: the typical urban home in industrialized countries wastes around 30% of food they buy, and the average closet has clothes that have been worn less than seven times before being discarded. The Boston Consulting Group estimates that the average buyer buys 60% more pieces of clothing than they used to in the year 2000, but retains each item for half the time.

"Our economy has gotten extremely good at making things cheaply, but extremely bad at getting them to places where they actually generate real well-being," observes economist Dr. Kate Raworth, creator of the "Doughnut Economics" framework.

This over-consumption has a twofold economic cost. First, we pay more than we need to for products that provide lower and lower returns in well-being. Second, as a group, we bear the environmental and social externalities, the non-market expenses not included in the prices but passed through to society through pollution, resource degradation, global warming, and waste disposal.

Studies repeatedly indicate that after the satisfaction of minimum needs, more consumption is not well correlated with greater well-being. A classic study in the Journal of Economic Psychology revealed that experiential spending repeatedly brought more happiness than material purchases of similar value. Likewise, time-saving expenditures often bring more satisfaction than status-seeking consumption.

When we look at our own spending habits with a critical eye, many of us find a great deal of waste, both money and material. Realizing this presents opportunity: by recognizing and removing wasteful consumption, we can both make our finances stronger and our footprint on the earth smaller.

Sustainable Investing and Personal Finance

Our money does work for us even when we're not, and more and more, folks want their money working for sustainability too. Sustainable investing has grown from a special interest to a mainstream financial planning strategy, and global sustainable investments have reached over $35 trillion as of 2020.

For private investors, sustainable finance no longer means simply shunning objectionable industries. Environmental, Social, and Governance (ESG) funds rank firms on both sustainability performance as well as on traditional financials. Impact investing allocates funds to solutions specifically to environmental and social problems. Green bonds invest in projects that have quantifiable environmental impacts.

"Evidence increasingly indicates that sustainable investments beat out traditional investments both competitively and by limiting exposure to certain risks," says financial planner Lakshmi Venkataraman. "Companies not ready for climate regulation, scarcity of resources, or shifting consumer preferences are in for a rough ride financially."

Aside from investment portfolios, our banking decisions also have a direct influence on sustainability. Conventional banks tend to lend money to industrial pollution, deforestation, and other unsustainable activities. Ethical banks and credit unions, on the other hand, adhere to open lending practices and environmental criteria. The Global Alliance for Banking on Values indicates that values-based banks provide similar financial returns while channelling capital towards sustainable development.

Digital technology has levelled the playing field for sustainable finance, allowing even individuals with limited resources to access it. Apps such as Claro Money and other financial management platforms assist in finding sustainable financial products, while platforms such as Plum and Oval facilitate micro-investing in sustainable funds for beginners.

True Cost Accounting in Daily Decisions

When we spend ₹50 on a cotton t-shirt, is that really its cost? Not at all. The sticker price doesn't account for many "externalities", actual costs that are pushed off somewhere else instead of being paid by the producer or consumer directly.

True cost accounting makes an effort to put figures to these indirect costs. An in-depth analysis conducted by Trucost found that the 3,000 biggest firms globally are causing environmental damage of more than $2.2 trillion each year, efforts fully subsidised by society and future generations. Transposed to consumer products, true cost accounting indicates stark imbalances between market price and true costs.

Think about the simple tomato. A traditionally produced tomato can be bought in the market for ₹20, whereas an organic one is sold for ₹30. The pricing difference is apparent until we include externalities: water contamination with pesticides, loss of soil fertility, carbon release from transportation, and medical expenditures due to exposure to pesticides. Research by The Economics of Ecosystems and Biodiversity (TEEB) initiative puts the real price of conventional agriculture at 1.5 to 3 times the market value.

We can use genuine cost thinking in our daily choices. When weighing transportation alternatives, look beyond the cost of the ticket and include pollution expenses, infrastructure needs, and health effects. When considering where to live, include commuting time, energy efficiency, and proximity to amenities that minimize overall consumption requirements.

"The intention is not to make all purchasing decisions inordinately complicated," states environmental economist Vikram Patel, "but to build up awareness of significant hidden costs so they can have an impact on our most important decisions."

Saving Resources Saves Money: The Dual Benefit

The idea that sustainability has to cost more is one of the most enduring myths in sustainability.

In fact, a great deal of resource-saving measures brings immediate economic savings in addition to their environmental benefits.

Energy efficiency measures generally provide the most transparent cost savings. LEDs consume 75% less energy than incandescent bulbs but have a 25-fold longer lifespan, making double savings in electricity and replacement costs. As per the Bureau of Energy Efficiency, an average Indian household can save ₹1,000-1,500 per month by implementing simple efficiency measures in lighting, refrigeration, and air conditioning.

Water conservation similarly generates direct savings. Low-flow fixtures reduce water consumption by 25-60% with negligible impact on functionality. More significantly, they reduce energy costs for water heating, one of the largest household energy expenses. The payback period for water-efficient fixtures is typically under one year in most urban areas.

Transportation options provide especially large reductions in cost. The Automotive Research Association of India estimates the total cost of car operation (purchase, insurance, maintenance, fuel, and parking) to be ₹25-30 per kilometre. Public transport costs about ₹2-5 per kilometre, and cycling costs paise per kilometre while delivering health benefits that save additional healthcare costs.

Even dietary decisions double as a double blessing. The Indian Council of Agricultural Research conducted a study that discovered households embracing plant-forward diets saved an average of 23% on food costs while cutting their carbon emissions by 30%.

"The most sustainable behaviors are usually those that avoid waste instead of needing new purchases," says consumer educator Anjali Desai. "Making use of what you already own in an efficient manner always saves money compared to purchasing something new, regardless of how 'green' the new item professes to be."

Building Wealth Through Conscious Consumption

Beyond immediate savings, conscious consumption creates pathways to longer-term financial security. The principles of wealth-building and sustainability share remarkable alignment, both favour durability over disposability, quality over quantity, and thoughtful planning over impulsive action.

The most straightforward link is straightforward: less consumption means more savings. Americans who embrace minimalist values claim to save 25-35% of their income versus the national average of less than 5%. These savings, when invested, generate compound returns that far exceed the fleeting pleasure of instant gratification.

Emphasizing strong, high-quality products also creates financial stability. Researchers at the Journal of Consumer Research discovered that customers who bought high-end versions of products that they used regularly saved 30% over five years from not constantly buying cheap, short-lived substitutes. The "buy once, buy well" strategy lowers lifetime costs while providing better user experiences.

"Sustainability-conscious consumers would intuitively lean toward the idea of 'enough,'" says financial planner Rohan Joshi. "They're able to differentiate between needs and wants better, establishing natural budgetary limits that keep capital intact for things that really count."

This attitude counteracts the "hedonic treadmill", our ability to quickly acclimatize to new levels of consumption, necessitating increasingly higher spending to keep up with the same level of satisfaction. In being satisfied with adequate rather than optimal consumption, conscious consumers build financial and psychic space for living on purpose.

Budgeting for Sustainability: Short-Term Costs vs. Long-Term Savings

Some sustainable options take more up front to achieve ultimate savings. Well-planned budgeting for sustainability involves reconciling these time-frames carefully.

Solar power illustrates this reality. It takes huge initial investment to put up rooftop solar panels, Rs. 1-1.5 lakh for a simple residential system in India, for instance. Yet the system produces free power for 25+ years, with payback times of 4-7 years based on sunshine available in the area and the rate of electricity. The long-term financial benefit is obvious, though one needs to plan for finances to absorb the initial burden.

The same considerations apply to home energy retrofits. Improving insulation, replacing leaky appliances, and incorporating smart energy management systems involve initial capital in order to harvest ongoing savings. Lifecycle cost analysis most often demonstrates 15-25% annual returns on these investments, better than most financial instruments, but the payback comes incrementally rather than instantly.

A variety of funding strategies can close this gap. Green loans provide low-cost terms for environmentally friendly home upgrades, and energy service companies sell performance contracts in which upgrades are financed by guaranteed savings from energy. Pay-as-you-save programs permit families to invest in efficiency at no up-front cost, sharing the accompanying savings with the provider instead.

For those with limited financial flexibility, incremental approaches work best. Start with no-cost and low-cost improvements: adjusting thermostats, optimizing existing systems, and eliminating waste. Use the resulting savings to fund higher-impact investments in a strategic sequence based on return on investment.

The Hidden Subsidies of Unsustainable Practices

Our economic system involves several subsidies that are embedded in nature and artificially lower the cost of harmful environmental activities and make sustainable options seem quite costly. Understanding these distortions enables us to make better decisions and pressure systemic reform.

Agricultural subsidies also mislead food markets to the detriment of sustainable agriculture. Subsidies generally go toward large mono-cultures of commodity crops and not diverse, regenerative agriculture. This artificially suppresses the cost of processed food compared to whole food, bringing about both environmental and health issues.

Waste disposal subsidies represent another hidden market distortion. Most municipalities charge flat fees for waste collection regardless of volume, effectively subsidizing high-waste households. Where waste charges reflect actual disposal costs, households typically reduce waste generation by 20-30%.

"These subsidies create a doubly perverse outcome," notes policy analyst Priya Mehta. "We pay once through our taxes to subsidize unsustainable systems, then pay again through environmental and health damages those systems create."

Knowing these distortions allows us to see that a lot of sustainable options are cost-effective even when they seem to cost more. The cost gap usually indicates artificially repressed costs for traditional options and not higher prices for sustainable options.

Call to Action

Monitor all "convenience spending" for one month and add up both the money cost and the resource cost, then determine reduction targets.

As we wrap up this chapter, here is a useful exercise to bring your financial and environmental objectives into harmony:

Step 1: Monitor Your Convenience Spending

- For one month, keep a record of all convenience-driven expenses: take-out meals, disposable products, expedited delivery charges, and tasks you can feasibly do yourself.
- Record both the financial cost and the inputs consumed (packaging, transportation, etc.).

- Use a plain notebook or an app like Expense Manager or Money Manager.

Step 2: Calculate Your Convenience Footprint

- Add up your monthly convenience spending, you might be amazed at the total.
- Estimate the yearly financial effect by multiplying by 12.
- Record the main resource effects: plastic waste, transport emissions, unnecessary packaging.

Step 3: Set Thoughtful Reduction Targets

- Pinpoint "low-hanging fruit", conveniences that offer little value compared to their expense.
- Establish clear reduction targets (e.g., "Cook at home 3 more times a month" or "Avoid express delivery charges").
- Invest some of your savings in higher-quality, longer-lasting goods or experiences that bring more well-being.

What convenience spending can you cut back on without lowering quality of life?

Chapter 9

Sustainability in the Digital Age

Technological Solutions to Resource Management

The same digital revolution that has revolutionized the way we communicate, work, and play is now transforming the way we relate to resources. Intelligent technologies are providing unprecedented possibilities to track, analyse, and optimize the use of energy, water, materials, and other finite resources.

Smart grids are likely the most revolutionary resource management technology. In contrast to traditional power grids that merely supply electricity in a single direction, smart grids establish a two-way exchange of electricity and information. This facilitates demand-response systems that automatically adapt consumption according to grid conditions, moving non-critical use to periods when renewable energy is plentiful. The World Economic Forum calculates that complete implementation of smart grids could cut carbon emissions from electricity by 12-18% and enhance reliability.

Water management has also changed with digital innovation. Smart water meters identify leaks immediately instead of after months of loss. Pilot projects with smart water infrastructure in cities such as Jaipur and Pune have cut water losses by 20% just by identifying leaks early. Irrigation systems for agriculture, the biggest water user in the world, can save 30-60% of water by providing exactly what the crop requires when it needs it using IoT and real-time weather and soil moisture readings.

Waste management systems today include digital intelligence along the value chain. RFID tags allow materials to be traced from production to disposal, enhancing recycling rates and informing circular economy efforts. Smart waste bins in cities such as Bengaluru and Chennai utilize fill-level sensors to maximize collection routes, cutting fuel consumption and vehicle emissions by up to 40% while avoiding overflow.

Building management systems are yet another resource optimization frontier. Office buildings with integrated digital systems commonly save 15-25% of energy through adaptive lighting, smart HVAC control, and occupancy-based adjustment. Extended to residential buildings using smart home technology, such systems enable households to cut waste without compromising comfort.

The Environmental Impact of Digital Life

While digital technologies offer powerful sustainability tools, they come with their own environmental footprint, one that grows as our digital activities expand. Understanding these impacts allows us to make more informed choices about our digital lives.

The manufacturing phase creates the most significant environmental impact for most digital devices. Producing a single smartphone generates approximately 60-70 kg of CO_2 equivalent, roughly the same as driving 400 kilometres in an average car. Beyond carbon, device manufacturing requires substantial mineral extraction, including rare earth elements often mined under environmentally damaging conditions.

The use phase generates continuous energy requirements, although these differ radically depending on device type. Although individual smartphones use very little electricity, their cumulative effect is large as they charge half a billion devices every day. Energy-intensive devices such as desktop computers, video game consoles, and big televisions can draw 150-300 watts during actual use, the equivalent of several refrigerators operating at the same time.

Internet infrastructure itself is energy-intensive. The transmission networks, data centres, and servers that power our online existence together use 1-2% of the world's electricity. Without ongoing efficiency gains, this could rise to 8% by 2030, a study in the journal Challenges said.

E-waste is likely the most visible environmental problem. We produce more than 50 million metric tons of electronic waste worldwide every year, with only 20% officially recycled. The rest usually ends up in landfills or unofficial recycling facilities that poison soil and water with toxins. A smartphone holds dozens of valuable materials that can be recovered via proper recycling, such as gold, silver, copper, and rare earth elements.

"The environmental impact of digital technologies is not inevitable but rather a consequence of design choices," notes Dr. Josh Lepawsky, author of "Reassembling Rubbish: Worlding Electronic Waste." "Through conscious design for longevity, energy efficiency, and end-of-life management, we can significantly reduce the environmental burden of our digital lives."

Apps and Tools for Sustainable Living

A new generation of apps is placing sustainability at our fingertips, facilitating the translation of environmental consciousness into real-world action through digital tools for daily use.

Carbon footprint calculators have moved beyond basic questionnaires to advanced tracking tools. Programs such as Earth Hero and Carbon

Footprint give users detailed breakdowns of individual emissions in transportation, home energy use, food, and consumption. The most useful versions create baselines, recommend targeted reductions, and monitor progress over time, making intangible carbon abstractions into concrete insights.

Sustainable shopping apps guide complicated purchasing choices. Fashion brands are graded by Good on You according to their ecological and ethical considerations, and scanning barcodes in products with Code Check provides disclosure of ingredients, sustainable packaging, and health effects. These platforms turn information asymmetry into something positive, where consumers are given transparency at purchase time.

Food waste prevention apps solve one of our biggest environmental problems. Too Good to Go links consumers with restaurants and stores selling excess food at reduced prices, while Olio enables food sharing between neighbors. Karma, which launched in India in 2022, has prevented more than 7,000 tons of food waste via its platform.

Transportation optimizers lower emissions as well as expense. Carbon footprint comparisons for alternative routes and travel modes are now offered by apps such as Citymapper and Google Maps, and carpooling apps such as Quick Ride and BlaBlaCar assist in filling otherwise empty seats on already traveling cars.

Energy management devices take utility-scale monitoring and provide it to individual homes. Smart meter-linked apps display electricity uses in real-time, and they identify energy guzzlers and propose ways for optimization. Research indicates that homes using these devices generally cut energy consumption by 7-12% based solely on behaviour change.

"Good sustainability apps don't just tell people something, instead, they turn habitual choices into chances for good," says digital sustainability specialist Maya Patel. "They work if they make sustainable options easier, more convenient, or more rewarding than the unsustainable alternative."

Social Media as a Force for Environmental Change

Aside from niche uses, mainstream social media have become strong drivers of environmental consciousness and action, linking people to causes, amplifying voices, and speeding up the diffusion of sustainable behaviors.

Millions for environmental causes all across have been mobilized by hashtag activism. Rising viral momentum in 2019, campaigns such as #TrashTag—which urged people to pick up trash and publish before-and-after images on social media—inspired thousands of cleaning events spread over several nations (BBC, 2019). Likewise, Rebecca Prince-Ruiz and the Plastic Free Foundation started #PlasticFreeJuly in Australia in 2011; from a local project, it has evolved into a worldwide movement involving millions of participants yearly in lower single-use plastic usage (Plastic Free Foundation, 2023).

Visual platforms like Instagram have mainstreamed eco-conscious lifestyles through aspirational content. Sustainability influencers demonstrate that environmentally responsible living can be aesthetically appealing and personally rewarding rather than a sacrifice. This visibility helps normalize practices that might otherwise remain niche.

Peer learning networks thrive on social media, where people exchange hands-on sustainability advice, solve problems, and support each other. Communities of practice such as zero-waste living, plant-based recipes, sustainable fashion, and the like form around these topics, speeding up adoption of environmentally friendly practices.

Targeted messaging engages once-disengaged people. While some environmental communications typically only reached persons already concerned, social media platforms' algorithms make sustainable content visible to diverse users based on near-interests. A fitness page follower may become aware of plant-based cooking recipes, while a home decor aficionado comes across up-cycled projects.

"As a social medium's most important sustainability contribution is a change in social norms," adds digital anthropologist Dr. Ananya Sen. "As we repeatedly observe people in our communities making sustainable lifestyle choices, they become from 'alternative' to 'normal' in our own minds, decreasing the psychological impediments to embrace."

Case Study: Tech Companies Leading in Sustainability

The digital industry creates powerful environmental effects but also hosts some of the most ambitious environmental efforts, illustrating how digital industries can reduce their impact while creating solutions for larger environmental issues.

Infosys became carbon neutral in 2020, well ahead of global climate targets. Its strategy blended large-scale renewable energy purchases, building and data centre energy efficiency, and carbon offset projects that targeted community benefits. Its Mysuru campus boasts India's largest corporate rainwater harvesting system, harvesting 25 million liters every year.

Microsoft has gone from carbon neutrality to

carbon negativity, pledging to offset all the carbon the company has emitted since its inception by 2050. Their sustainability plan features an internal carbon charge that places a cost on emissions for every business unit, generating financial incentives for cutting back. The company's AI for Earth initiative offers cloud computing capacity and artificial intelligence tools to organizations tackling environmental issues.

Apple's round production efforts show how technology firms can challenge resource constraints. Their recycling machine, Daisy, can break down 1.2 million iPhones every year, reclaiming materials to be used in new products. The firm has also phased out plastic packaging for most lines and pledged to 100% carbon-neutral products by 2030.

Smaller players have led sustainability innovation for the digital world. Fairphone develops phones designed for longevity, repairability, and ultimate recycling, while Framework develops modular, upgradeable laptops that increase useful life and lower e-waste.

"These leaders show that sustainability isn't a secondary aspect of business strategy but fundamental to long-term resilience," says corporate sustainability analyst Vikram Menon. "They're already planning for resource scarcity, changing regulations, and shifting consumer attitudes that will transform markets over decades to come."

The Carbon Footprint of Cloud Computing and Digital Services

The seemingly weightless world of digital services, streaming media, cloud storage, online gaming, videoconferencing, has very real physical infrastructure with corresponding environmental impacts. Understanding this footprint helps us make more informed choices about our digital consumption.

Data centres form the foundation of cloud services, consuming approximately 1% of global electricity. While this figure has remained relatively stable despite exponential growth in data processing, it represents significant absolute energy use. The most efficient hyperscale data centers use 50% less energy per computational unit than typical enterprise data centers, highlighting the importance of infrastructure design.

Video streaming is the biggest group of internet traffic and related energy consumption. High-definition video streaming for one hour takes around 0.2-0.3 kg of CO_2 depending on device type and carbon intensity of power grid. Although it may appear small per individual, as a cumulative group, video streaming is responsible for about 1% of worldwide greenhouse gas emissions.

The environmental footprint differs significantly between digital services. A basic email has a negligible footprint (0.3-4g of CO_2), whereas a video call produces about 150-1,000 times as much carbon per minute. Cloud storage services also have very different environmental footprints depending on data centre efficiency, use of renewable energy, and data redundancy strategies.

Cryptocurrency and blockchain technologies pose specific issues. Bitcoin alone uses more electricity than numerous nations because its energy-hungry "proof of work" consensus mechanism is so power-intensive. Other methods such as "proof of stake" cut energy needs by over 99%, illustrating that the environmental footprint of blockchain is a matter of design rather than an intrinsic constraint.

"The critical observation is not that we ought not use digital services but rather that we must be responsible about the manner in which we utilize them," notes digital sustainability researcher Priya Mehta. "Low-impact habits such as downgrading video quality where audio content dominates, erasing unwanted stored data, and opting for service providers who are known for environmental integrity can cut significantly on our digital impact."

Utilizing Data Analytics to Enhance Personal Sustainability

The same analytical methodologies that inform business for optimal operation can rework personal environmental footprint if brought to bear on individual consumption behaviour. Decision-making driven by data delivers precision and efficacy to sustainability.

Smart home systems are the most direct type of personal sustainability analytics. By linking information from smart thermostats, energy monitors, water sensors, and other devices, homes receive unparalleled insight into consumption patterns. The most advanced systems detect anomalies (such as water leaks or faulty appliances), recommend optimization techniques, and measure behaviour change savings.

Personal carbon accounting software uses business-grade analytics on personal footprints. Mobile apps such as Joro and CarbonClick monitor expenditure by categories, add carbon intensity multipliers to each transaction, and create personal "carbon statements" and financial overviews. It renders carbon real by relating it to concrete everyday decisions instead of guesstimates.

Feedback loops speed up improvement in any system. The best personal sustainability tools form tight feedback loops between behaviour and outcome, providing instantaneous results from changes in behaviour. This psychological feedback improves the formation and maintenance of habits more than feedback that is delayed or vague.

Predictive suggestions are the future of personal sustainability analysis. By looking at trends across large groups of people, sophisticated systems can offer very targeted recommendations with maximum potential benefit per person. Instead of one-size-fits-all suggestions to lower energy use, these systems may prescribe specific things to do based on home type, local weather, and observed behaviour patterns.

"Personalization is the solution to successful sustainability action," observes behavioral data scientist Dr. Anil Gupta. "Generic recommendations are easy to write off as irrelevant to our particular case. Data analytics takes away this excuse by making recommendations exactly fitting to our individual circumstances and consumption behaviors."

Call to Action

Adopt one digital tool to monitor and lower your ecological impact, and pass it on to friends.

As we close out this chapter, here's a step to apply digital technology to sustainability in practice:

Step 1: Pick Your Tool

Pick a digital tool consistent with your environmental focus:

- **For carbon consciousness:** Carbon footprint calculator (Joro, Carbon Footprint)
- **For energy conservation:** Smart home energy monitor (Sense, Smappee)
- **For transport:** Transit or carpool app (Citymapper, Quick Ride)
- **For consumption:** Sustainable shopping guide (Good on You, Code Check)
- **For food waste:** Food sharing platform (Olio, Too Good to Go)

Step 2: Set Your Baseline

- Measure your existing impact in the area you've chosen using your selected tool.
- Record important metrics as baselines (e.g., monthly carbon footprint, daily energy consumption).
- Determine the most impactful areas for improvement.

Step 3: Share Your Progress

- Share your sustainability tool on social media.

- Encourage friends to join you in using the app.
- Post regular updates on your progress and learning.
- Celebrate milestones and improvements, creating positive reinforcement.

Which digital sustainability tool fits best with your lifestyle and priorities?

Chapter 10

From Individual to Collective: Scaling Sustainable Solutions

Community-Based Sustainability Initiatives

Individual actions build the base of sustainable living, but community-based initiatives multiply effects and drive systemic change that any individual could not achieve alone. These local efforts illustrate how collective action can reshape neighbourhoods, towns, and cities into laboratories for sustainable innovation.

Transition Towns is one of the most widely used community models of sustainability. Initiated in Totnes, UK, and now covering more than 1,000 communities across 50+ nations, the Transition approach is centred on creating local resilience to climate change and resource limitation. Every Transition project is responsive to local conditions, but they share some features such as local renewable energy projects, local food systems, networks of skill exchange, and local currencies that help keep money within the community.

Urban agriculture and community garden projects convert vacant land into productive landscapes and foster social ties. Mumbai's Urban Leaves initiative has established more than 60 community gardens on rooftops and vacant plots, yielding fresh produce while mitigating urban heat islands and enhancing air quality. The areas become living classrooms where sustainable principles are taught across generations and social divides.

Tool libraries and repair cafes represent the sharing economy on the community level. Delhi Tool Library enables members to borrow costly equipment instead of buying seldom-used pieces of equipment, while Bengaluru hosts a monthly Repair Café wherein volunteers with technical expertise come together to repair broken items that are otherwise destined for the landfill. These models lower consumption levels while promoting community relations and resiliency.

Renewable energy owned by the community shows the strength of mass investment. European energy cooperatives now own some 7% of installed renewable energy capacity, while Indian solar cooperatives are powering three million rural homes that the traditional grid neglected. These actions democratize ownership of energy as it speeds the shift to renewables.

"Communities are the laboratories where sustainable solutions are refined before broader adoption," notes social ecologist Dr. Amrita Sen. "They're small enough to implement changes quickly but large enough to demonstrate viability at meaningful scale."

Corporate Responsibility and Consumer Advocacy

The business sector wields enormous influence over global resource flows, supply chains, and consumption patterns. Harnessing this power for sustainability requires both internal corporate leadership and external pressure from consumers and investors.

Corporate sustainability has developed dramatically during the past decades. The first generation of strategies was mainly aimed at compliance and risk management, doing less harm through pollution reduction and waste management. The second generation moved into resource efficiency and product innovation. Regenerative business models that work to restore natural systems, rather than merely avoiding damage, are being adopted by today's industry leaders.

Consumer activism has played a key role in pushing this transformation. Organized campaigns have revolutionized entire sectors, from sustainable seafood labelling to non-toxic electronics. The #BeatPlasticPollution campaign led large food and beverage brands to rethink packaging after millions of customers indicated their priorities through shopping choices and direct action.

The best advocacy takes into account the strength of market signals. When Greenpeace called attention to deforestation in the palm oil industry, they targeted large consumer brands instead of unknown producers. This approach used brand reputation issues to create swift changes across the supply chain.

Digital platforms have democratized consumer advocacy. Apps like Buycott allow shoppers to scan product barcodes to see if items align with their values, while platforms like Change.org and Avaaz enable rapid mobilization around specific corporate practices. Social media campaigns can quickly elevate issues to boardroom attention through viral sharing.

"Consumer advocacy is most effective when it marries precise, concrete demands with acknowledgment of progress," says campaign strategist Rohan Joshi. "Companies must understand both what is anticipated to change and that the changes will be noticed when made.

While individual decisions are important, their impact grows exponentially when they are empowered by policies. The proper regulatory systems can eliminate obstacles, create incentives, and create an environment where sustainable alternatives are the easiest and lowest-cost alternatives.

Building codes demonstrate this principle well.

When governments adopt updated codes calling for improved insulation, efficient appliances, and solar-readiness, they make it a requirement for all new homes to be more advanced. In doing so, they prevent leaving the responsibility for sustainable decisions with individual home buyers who might not have technical information or bargaining strength.

Waste management policy also influences individual behaviour. Pay-As-You-Throw communities, where residents pay for the waste they produce, generally cut waste by 25-45% in the first year of operation. Such policies align economic incentives with environmental objectives, making sustainable options economically preferable.

Infrastructure for transport decides every day's trip decisions more strongly than individual choice. Those cities that spend money on safe cycling networks, good public transit, and foot-friendly streets find people naturally change to sustainable transport. Copenhagen's planned investment in cycling facilities has led to 62% of citizens commuting by bike, not because Danes are "greener" people by nature, but policy made the bike the most convenient means of transport.

Extended Producer Responsibility (EPR) policies move the cost of waste management from taxpayers to manufacturers, providing the incentive to create for durability, repairability, and recyclability.

When businesses are forced to account for their products' end-of-life burdens, they will inevitably redesign them in order to reduce these burdens.

"The most effective policies make sustainable choices the default, rather than making them something that requires conscious effort," says policy analyst Vikram Mehta. "They design systems where the right thing is the path of least resistance."

Global Networks, Local Impact

Environmental issues transcend all boundaries, demanding geographically and culturally coordinated responses. A new global ecosystem of networks is forming to enable knowledge exchange, resource mobilization, and coordinated action while being sensitive to local contexts and leadership.

City networks have proven especially adept at scaling up sustainability innovations. The C40 Cities Climate Leadership Group operates among 97 global cities that have a population of 700+ million and account for 25% of the world economy. Successful policies such as bus rapid transit systems and building energy standards have diffused across cities quite rapidly via this network, speeding up implementation through peer-to-peer knowledge transfer and friendly competition.

The Global Ecovillage Network is connected to thousands of intentional communities and traditional villages that are showing sustainable living in action. These living labs experiment and refine regenerative strategies in agriculture, building, energy, and social organization, producing models that can be replicated in a variety of contexts.

Financial systems channel capital into sustainability solutions that would otherwise remain unfunded. The Global Alliance for Banking on Values brings together 67 financial institutions with 77 million customers and more than $210 billion in assets under management, all dedicated to employing finance to generate positive social and environmental impacts.

Digital platforms enable global-local linkages without the need for physical proximity. The Sustainability Open Innovation Platform links problem-holders and solution-providers on different continents, while blockchain technologies allow verification of environmental claims along global supply chains.

"The best global networks understand that solutions need to be tailored to local contexts, not applied in a one-size-fits-all manner," says international development specialist Lakshmi Narayan. "They offer frameworks, knowledge, and resources while empowering local leadership and decision-making."

Building Sustainable Change: Sustainability as Cultural Norm

Sustainability must become part of cultural norms, the unspoken rules that govern everyday behaviour and social expectations in order to shift from conscientious choice to default action. A number of avenues enable this cultural integration.

Education systems are responsible for embedding sustainability across generations. Once environmental literacy is made part of curricula and not an afterthought, students learn ecological knowledge as a foundational decision-making framework. The best methods embed sustainability across subject areas instead of addressing it as a standalone subject.

Media and entertainment influence cultural narratives significantly. When sustainable behaviours are represented as the norm and something to be desired in movies, TV, and social media, they become aspirational, not sacrificial. This shift in representation assists in overcoming the perception that living sustainably means living in deprivation.

Institutional habits set strong behavioural signals. When schools, workplaces, and community organizations turn to sustainable practice, from green buildings to plant-focused cafeteria menus, their environments allow sustainable choices to become the route of least resistance.

Framing and language shape our understanding of sustainability. Moving away from technical language and toward values-based framing ties environmental actions to commonly held principles such as responsibility, stewardship, and concern for future generations. This framing bridges political and social divides that could otherwise prevent collective action.

"Cultural change happens when a number of influences converge to support new patterns of thought and behaviour," observes cultural anthropologist Dr. Maya Patel. "The transition to sustainability gains momentum when education, media, institutions, and social networks all communicate that environmental stewardship is just 'how we do things here."

Intergenerational Equity and Long-Term Thinking

Perhaps the biggest challenge in sustainability is widening our circle of moral concern to include future generations. Though most ethical theories acknowledge our duties to contemporaries, our duty to those who will inherit the earth is underrepresented in decision-making.

Intergenerational equity, the idea that current generations must not sacrifice the capacity of future generations to satisfy their own needs, demands a fundamental change in temporal orientation. Political and economic institutions in most societies maximize short-term returns, and therefore there is a structural bias against investments whose payoffs accrue mainly to future citizens.

Indigenous wisdom traditions provide good models for long-term thinking. The "seventh generation principle" of the Haudenosaunee (Iroquois) Confederacy teaches leaders to think about how a decision will impact descendants seven generations later, about 150 years. This approach naturally connects decision-making with sustainability principles.

Institutional innovations can assist in the integration of future interests into current choices. Some jurisdictions have created "Guardians for Future Generations" with powers to scrutinize legislation for long-term effects. Others have incorporated environmental rights into constitutions, giving rise to legal standing for intergenerational interests.

Personal practices can foster long-term thinking. Becoming connected with nature through gardening, forest preservation, or planting trees makes us literally invested in future time periods. Working on tasks whose fruits will survive us, whether rehabilitating ecosystems or constructing community institutions, elongates our temporal perspective.

"The heart of sustainability is to understand that we are holding the natural world in trust, not possession," says environmental philosopher Dr. Sunita Sharma. "We are borrowing the planet from our children and not inheriting it from our parents."

Building Resilient Systems for Environmental Uncertainty

As the environment becomes more and more unpredictable, resilience, the ability to resist, recover from, and adapt to disturbances, becomes crucial. Designing resilient systems involves particular design principles that are different from the conventional efficiency-oriented approach.

Diversity is an important resilience mechanism. In farming, polycultures are more resilient to pests and weather shocks than monocultures. In energy systems, diverse generation sources and storage technologies offer reliability in the face of disruptions. In communities, diverse skills and knowledge facilitate adaptable responses to challenges.

Redundancy, frequently cut back in efficiency-based systems, creates vital backup capacity during outages. Seed banks save farm choices for shifting conditions, and microgrids sustain power to key services when main grid failures occur. These so-called "inefficient" copies become indispensable when primary systems fail.

Modularity, designing systems so that components can operate independently when needed, improves resilience by isolating failures instead of letting them cascade. Distributed renewable energy systems illustrate this principle, still delivering power in some regions even when others are hit by extreme weather.

Feedback mechanisms allow for early warning and response to shifting conditions. Environmental monitoring networks, citizen science, and traditional ecological knowledge all offer useful information that can activate adaptive responses prior to crossing thresholds.

"Resilience thinking is a radical departure from presuming stability to accepting change," says systems ecologist Dr. Rajiv Kumar. "Instead of attempting to hold fixed conditions, resilient systems continuously adapt while maintaining key functions."

Call to Action

Find one issue that you care deeply about and take a concrete step to advocate for systemic change, write a company, call on a legislator, or become involved in an organization addressing the issue.

As we reach the end of this chapter and this book, here's a final call to action that stitches together individual practice and systemic change:

Step 1: Identify Your Issue

- Select a sustainability challenge that resonates intensely with you personally.
- Consider local environmental issues directly affecting your community.
- Focus on areas where your knowledge, skills, or position provide unique leverage.
- Research existing efforts addressing this issue to avoid duplicating work.

Step 2: Determine Your Intervention Point

Identify which systems influence your chosen issue:

- Is it primarily a business practice that needs changing?
- Does it require policy intervention at local, state, or national levels?
- Would community mobilization create the necessary pressure?
- Think about where your voice would be most powerful based on your position and networks.

Step 3: Take One Specific Action

- **For corporate change:** Email a company regarding particular practices, with particular requests.

- **For policy change:** Call a legislator with a specific ask based on your cause.
- **For community change:** Volunteer with an organization tackling your cause, sharing particular skills.
- **For knowledge change:** Host an educational event to create awareness and commitment.

Which sustainability challenge will you champion, and what concrete action will you take this month?

Conclusion

A Sustainable World Within Reach

Throughout this book, we've examined the various aspects of sustainability in everyday life. From the tiny steps that minimize waste to the bigger systems that determine our combined effect, we've discovered that making a sustainable world is not impossible or utopian but practical, worthwhile, and increasingly vital.

We started by redefining sustainability from buzzwords to something that acknowledges our smallest actions can have ripples of good. We looked at our relationship with material things, challenging the economics of excess and finding the freedom in "enough." We thought about how simple changes in attitude can change communities and how smartly managing resources such as energy and water is good for our planet and our wallets.

We faced problems such as noise pollution and microplastics which tend to go unnoticed but take a major toll on our wellbeing. We considered how the convenience economy could be reimagined to work for us without straining the planet, and how our economic choices relate to ecological outcomes. We looked at digital technologies that have the power to speed up sustainable behaviours, and lastly, how we might turn individual action into collective action.

Throughout this journey, several key insights have emerged:

First, sustainability isn't about sacrifice, it's about improvement. The most enduring sustainable practices enhance our lives rather than diminishing them. Energy-efficient homes are more comfortable. Plant-rich diets improve health. Reduced clutter creates mental space. Walking and cycling build community connections while reducing pollution. When properly implemented, sustainable choices deliver multiple benefits that extend far beyond environmental impact.

Second, small things do count, particularly when they become habits and the norm. The compounding effect of millions of individuals making marginally better decisions generates massive positive impact. No one can do it on their own, but neither can it be done without it happening at the individual level. Our everyday decisions are the building blocks of big solutions.

Third, both systems and people need to co-evolve. Individual responsibility works within systems capable of supporting or sabotaging choices that are more sustainable. Instead of arguing if change needs to be top down or bottom up, we want both: better choices by people while pushing systems that make people's better choices easier for everybody.

Fourth, sustainability is a practice, not an end point. There is no finish line of sustainability to step across and never look back. Instead, sustainability is an ongoing journey of learning and improvement. This is not a cause for dismay but for humility and continued participation.

Finally, sustainability is about connection, with nature, with each other, and with the future. Sustainable living at its core is about recognizing our interdependence and living in response. It is about understanding that our health is inextricably tied to the health of the living systems that support us.

As you finish this book and press on with your own sustainability path, keep in mind that you're not by yourself. Millions of individuals across the globe are making comparable decisions and facing comparable challenges. Together, we're developing new styles of living that are more satisfying, more just, and more aligned with the natural world.

The way forward is not always obvious or simple. There will be contradictions, setbacks, and hard choices. Progress is often a zigzag path. But the direction is certain: towards more awareness of what we are doing to the world, more responsibility in our actions, and more concern for the world we live in.

Whatever your next step, is installing a water-saving appliance, participating in a community garden, working towards policy change, or just becoming more conscious of what you buy, it counts. Every action stock your ability to take the next one and shows others what you can do.

A sustainable world isn't some future utopia. It's happening now, as a result of millions of daily decisions by folks like you. It arises out of the collective intelligence of communities discovering more ways to satisfy their needs without sacrificing natural systems. It develops as innovations bring prosperity with reduced material throughput and less environmental degradation.

The question is not if we can build a sustainable world, but whether we will opt to join in its construction. The offer is to each of us, and the moment to take it is now. The journey continues beyond these words, in the decisions we make and the future we construct together.

Your next decision is waiting. Make it matter.

Bibliography

1. **Brundtland Commission.** (1987). *Our Common Future: Report of the World Commission on Environment and Development*. Oxford University Press.

2. **Cialdini, R. B.** (2003). Crafting normative messages to protect the environment. *Current Directions in Psychological Science*, 12(4), 105-109.

3. **Duhigg, C.** (2012). *The Power of Habit: Why We Do What We Do in Life and Business.* Random House.

4. **Ellen MacArthur Foundation.** (2017). *A New Textiles Economy: Redesigning Fashion's Future*. Ellen MacArthur Foundation.

5. **Global Footprint Network.** (2021). *National Footprint and Biocapacity Accounts*. Retrieved from https://www.footprintnetwork.org/

6. **International Energy Agency.** (2021). *Energy Efficiency 2021*. OECD/IEA.

7. **IPCC.** (2022). *Climate Change 2022: Mitigation of Climate Change. Contribution of*

Working Group III to the Sixth Assessment Report of the Intergovernmental Panel on Climate Change. Cambridge University Press.

8. **Kasser, T.** (2002). *The High Price of Materialism*. MIT Press.

9. **Lepawsky, J.** (2018). *Reassembling Rubbish: Worlding Electronic Waste*. MIT Press.

10. **Raworth, K.** (2017). *Doughnut Economics: Seven Ways to Think Like a 21st-Century Economist*. Chelsea Green Publishing.

11. **Rochman, C. M., Tahir, A., Williams, S. L., Baxa, D. V., Lam, R., Miller, J. T., ... & Teh, S. J.** (2015). Anthropogenic debris in seafood: Plastic debris and fibers from textiles in fish and bivalves sold for human consumption. *Scientific Reports*, 5(1), 14340.

12. **Schultz, P. W., Nolan, J. M., Cialdini, R. B., Goldstein, N. J., & Griskevicius, V.** (2007). The constructive, destructive, and reconstructive power of social norms. *Psychological Science*, 18(5), 429-434.

13. **Thaler, R. H., & Sunstein, C. R.** (2008). *Nudge: Improving Decisions About Health, Wealth, and Happiness*. Yale University Press.

14. **United Nations Environment Programme.** (2019). *Global Environment Outlook – GEO-6: Healthy Planet, Healthy People*. Cambridge University Press.

15. **World Health Organization.** (2018). *Environmental Noise Guidelines for the European Region*. WHO Regional Office for Europe.

INDEX

A

Actionable insights – 6, 39, 147

Adaptation strategies – 25, 67, 88

Advertising psychology – 19, 34, 111

Air pollution – 58, 94, 133

Apps for sustainability – 87–89

Attitudinal shifts – 14, 42, 165

Awareness creation – 3, 12, 28

B

Behavior change – 11, 13, 22, 29

Biodegradable products – 54, 116

Biodiversity – 90, 117

Budgeting for sustainability – 79–82

Buy Nothing Project – 44, 52, 146

C

Carbon emissions – 5, 8, 33, 89

Circular economy – 48–50

Clothing & fashion waste – 27, 31–34

Community efforts – 52–55, 120

Composting – 45, 116

Consumerism – 2, 27–30

Cost-benefit analysis – 78–79

Cultural adaptation – 15, 52

D

Decision-making, sustainable – 12, 20, 91

Digital minimalism – 7, 86, 145

DIY solutions – 43, 116

Downshifting lifestyle – 41–42

E

Economic implications – 77–80, 154

Education & awareness – 24, 67, 104

E-waste – 35–38, 88

Energy efficiency – 20, 70–73

Environmental ethics – 13–16

Environmental myths – 17–21

F

Fast fashion – 32–34

Finance & sustainability – 77–82

Food habits – 5, 44, 99

Footprint calculators – 22, 89

G

Green apps – 87

Guilt-based motivation – 13, 21

H

Habits, personal – 10–11, 28, 114

Health, mental and physical – 13, 71

Household resource use – 3, 46, 92

I

India, practices in – 5, 7, 51

Individual responsibility – 4–5, 21

Innovation in sustainability – 90–91

L

Land use – 113–114

Laws & regulation – 137

Lifestyle transformation – 2–3, 43

Local movements – 50–53

M

Materialism – 27–29, 108

Mental models – 11, 30

Microplastics – 3, 18, 67

Minimalism – 40–42

Mindful living – 14–15, 30

Misinformation – 17–20

N

Natural resources – 5, 23, 72

Noise pollution – 70–72

O

Obsolescence, planned – 35–37

Overconsumption – 3, 27, 78

Ownership costs – 46–49

P

Packaging waste – 6, 74

Personal finance – 79

Plastic pollution – 4, 5, 67

Policy suggestions – 147–148

Pollution types – 58, 67, 70

Public participation – 51–53

Purchase habits – 27, 30

R

Recycling limits – 18–19, 38

Repair movement – 36, 43

Resource depletion – 45, 90

Reuse strategies – 27, 44

Risk perception – 10, 19

S

Self-sufficiency – 43, 116

Shared economy – 50–54

Shopping psychology – 28–30

Social behaviors – 11–12

Social media impact – 87

Sustainable investing – 79

Sustainability tools – 22–24, 87

T

Technology & consumption – 6–8, 86

Transportation – 5, 70, 97

Trash auditing – 19, 116

U

Urban living – 41, 67, 144

Usage patterns – 23–24

V

Values shift – 14–15, 26

Visual consumption cues – 13, 30

W

Waste management – 3–5, 18, 45

Water use – 24, 46, 91

Wealth and sustainability – 79–80

Made in the USA
Monee, IL
04 November 2026

67379243R00115